T5-CQJ-690

SUSTAINABLE COMMUNITY DEVELOPMENT SERIES

Chris Maser, Editor

Resolving Environmental Conflict:
Towards Sustainable Community Development

Sustainable Community Development:
Principles and Concepts

Setting the Stage for Sustainability:
A Citizen's Handbook

The Role of Vision and Leadership

SUSTAINABLE COMMUNITY DEVELOPMENT SERIES

Setting the Stage for SUSTAINABILITY

A Citizen's Handbook

Chris Maser • Russ Beaton • Kevin Smith

Lewis Publishers

Boca Raton Boston London New York Washington, D.C.

Library of Congress Cataloging-in-Publication Data

Catalog information may be obtained from the Library of Congress

Lewis Publishers is an imprint of CRC Press LLC

No claim to original U.S. Government works
International Standard Book Number 1-57444-187-6
Printed in the United States of America 1 2 3 4 5 6 7 8 9 0
Printed on acid-free paper

Go to the people.
Live with them.
Learn from them.
Love them.
Start with what they know.
Build with what they have.
And with the best leaders,
when the work is done,
the task accomplished,
the people will say,
we have done this ourselves.

Lao-tzu

The taproot of all civilizations to come lives in each of us in the present. To nurture the future is to honor the seeds of the present by allowing them to grow.

Jamie Sams

To the memory of my friend Bob Rodale, who began working toward the notion of social/environmental sustainability before I did. Although I never had the pleasure of meeting Bob face to face, our correspondence and phone conversations continue to inspire my efforts. Thank you, Bob.

Chris

I dedicate this book to my wife, Delana, and to our three children, Lynn, Alan, and Brenda. It is with the hope of helping to create a better world for them that I join Chris and Kevin in conveying our message.

Russ

I dedicate these words and ideas to the next generation, my two sons, Jason and Jeremy.

Kevin

TABLE OF CONTENTS

EDITOR'S NOTE

The book you are holding is part of a series on the various aspects of sustainable community development, where "community" focuses on the primacy and quality of relationships among people sharing a particular place and between people and their environment. "Development" means personal and social transformation to a higher level of consciousness and a greater responsibility to be one another's keepers, and "sustainability" is the act whereby one generation saves options by passing them to the next generation, which saves options by passing them to the next, and so on.

This series came about because, during the 25 years I was in scientific research, I discovered disturbing patterns of human thought and behavior that continually squelch sustainable community development. These patterns are as follows:

1. While physicists have found a greater voice for the spiritual underpinnings of physics, the biological sciences have all but lost their spiritual foundation, casting us adrift on a sea of arrogance and increasing spiritual, emotional, and intellectual isolation.
2. There is a continuing attempt to force specialization into ever-narrowing mental boxes, thereby so fragmenting our view of the world that we continually disarticulate the very processes that produce and maintain the viability of ecosystems on which we, as individuals and societies, depend for survival.
3. People point outside themselves to the cause of environmental problems without understanding that all such problems arise within ourselves, with our thinking. Before we can heal the environment, we must learn to heal ourselves emotionally and spiritually.

4. We are asking science to answer questions concerning social values, which science is not designed to do. Social questions require social answers.
5. One who has the courage to ask questions outside the *accepted* norm of scientific inquiry is ostracized because, as English philosopher John Locke said, "New opinions are always suspected, and usually opposed, without any other reason...[than] they are not already common."

This series of books on the various facets of sustainable community development is thus a forum in which those who dare to seek harmony and wholeness can struggle to integrate disciplines and balance the material world with the spiritual, the scientific with the social, and in so doing expose their vulnerabilities, human frailties, and hope, as well as their vision for a sustainable future.

Chris Maser
Series Editor

FOREWORD

As humans, we make choices. Many say this is what distinguishes us from other creatures. With change as a constant, we are continually presented with a great number of choices—and we must choose. The change represented by the divergence of humanity from the rest of the natural world is massive, rapid, and in need of transformation. This book, with its strong focus on the individual and her or his willingness to grow and forge new relationships with herself or himself, others, and her or his community, is a guide for that transformation, which can help create a sense of place where it may not now exist.

Until I moved to the Lake Superior Basin of northern Wisconsin, I did not understand that it is a sense of place that brings people into civic life, with its committees, commissions, and governmental decision-making process. I did not understand that a sense of place starts with the individual. Nor did I fully understand, even after 20 years as a city planner, that communities with this kind of heart are more likely to value planning as a tool of self-determination and therefore are less likely to allow market forces to shape their futures.

I know what market forces do. I live in a former boomtown, a shipping terminal for the natural resources of the area—lumber, ore, granite, and brownstone. When all of this natural wealth had been extracted, when its abundance was exhausted, the company left the "company town." We are still recovering—four decades later. Fortunately, we are now advancing new ways of thinking and positioning ourselves for a healthier future.

Setting the Stage for Sustainability: A Citizen's Handbook provides a rich and valuable understanding of how we can nurture a healthier future and cultivate a connection to one another and to the place in which we live. People typically develop a sense of "place" through investments (both material or financial and nonmaterial), such as work, family, groups and organizations of mutual interest, recreation, and so on.

Other kinds of connections are found through such actions as protecting spaces and buildings of historic, cultural, or spiritual value; initiating annual community-wide events; recognizing people and their contributions to the community; and integrating music and the arts into the daily life of the community. But, most importantly, before we can create social/environmental sustainability, we need to learn, understand, and embrace the reasons why so many people feel unrooted and apart in their own communities.

The authors have made many contributions in *Setting the Stage for Sustainability: A Citizen's Handbook.* Among them are positive and constructive ways of looking at potential obstacles to building a sustainable community. A major obstacle is citizen apathy, represented by low turnouts of voters and a general decline in participation in civic life, which have led many people to a skeptical if not cynical view of our democracy.

If no one shows up to vote, do we have a democracy? Such cynicism can be just as damaging to the achievement of the democratic ideal as apathy. Like fear, cynicism can paralyze.

Taking a positive view, however, the authors see apathy as "a disguise for a deep hunger to learn within the safety and nurturance of community." They choose to see this unexpressed power of the citizenry as fuel for change, rather than as waste, disarming the negative interpretation that would validate the cynic.

Maser, Beaton, and Smith say that through positive thinking and the willingness to risk, we can indeed be creative forces in our respective communities and in the world—converting societal waste into fuel or food for microorganisms, dysfunction into function, fear into hope and even excitement about the possibilities lurking in risk. They challenge us to be artists and embrace the full palette of opportunities that are at hand, to design a future, the exact dimensions of which we need to be comfortable not knowing, even as we begin to paint. They advocate making plans but not planning the results, believing that it is both the positive process and trust in the process that will foster the kind of world we all want to have and of which we want to be a part.

Consistent with the principles of sustainability, the authors analyze and describe good and bad institutionalized social patterns in an ecological sense. They emphasize the importance of each component to the whole, in this case the individual person.

If we want to be proponents of biodiversity, we must honor individuals and their unique contributions to the ecosystem of which we are all an inseparable part. We must see these contributions as essential ingredients in the integrity and healthy functioning of the whole. But individuals must value their own contributions as well, something that community, government, organizations, business, and industry can help engender by "empower[ing] individual intelligence and honor[ing] intuition."

By helping us to see the connection of the individual to the whole, the one to the many, this book stands out as a remarkable tool, as does *Sustainable Community Development: Principles and Concepts*. Each invites us to begin setting the stage for sustainable communities, ones more in harmony with the rest of the natural world, where understanding and acceptance of the current situation form the first step toward change.

The authors guide us toward an understanding and acceptance of our social failure, particularly in the United States, to adapt to the natural landscape in ways that perpetuate life and conserve options for future generations through a thoughtful analysis of our economy, its history, and its evolution. In addition, Maser, Beaton, and Smith help us to see conflict as a natural occurrence, something inherent in any natural system and indeed a path to truth if accepted and *used*. Basic principles of conflict resolution are set forth, guided by an understanding of conflict as fuel for change.

Above all, the importance of questions is illuminated. "People do not grow by knowing all the answers; they grow by living with the questions and their possibilities." The message inspires us to value our curiosity, intuition, imagination, and ignorance as essential tools with which to see the possibility of a world lean on rule, steadfast in personal commitment, and filled with the courage to create artful and authentic lives, where all of us are relevant and valued parts of the whole, where we can inspire others toward embracing words like those of Mary Oliver:

> Whoever you are, no matter how lonely,
> the world offers itself to your imagination,
> calls to you like the wild geese, harsh and exiting—
> over and over announcing your place
> in the family of things.

Jane M. Silberstein
Washburn, Wisconsin

Ms. Silberstein was for 15 years a city planner, first in Santa Barbara, California, and then in Santa Cruz, California. Today she is the U.S. Coordinator for the Lake Superior Binational Forum to protect and restore Lake Superior. In addition, Ms. Silberstein continues to work as a consultant in community planning and development in northern Wisconsin.

PREFACE

This book, *Setting the Stage for Sustainability: A Citizen's Handbook,* is the third in a series dealing with sustainable community development. The first book, *Resolving Environmental Conflict: Towards Sustainable Community Development,* is a necessary point of departure for the journey, because destructive environmental conflicts are, above all, intergenerational, and their effects, which compound over time, often destroy potential sustainability.

Environmental conflicts must be resolved through the "transformative" approach to facilitation because the sustainability of one's community within the context (from the Latin *contexere,* which means to weave together) of its landscape requires personal growth through a shift in consciousness and a willingness to let go of self-centeredness in favor of other-centeredness—the core of a shared vision of the future toward which to build. Facilitation, as it is here used, means to conduct a process of communication whereby people are assisted in freeing themselves from difficulties and obstacles in making decisions that either avoid or eliminate destructive conflict by forging commonly held values into a shared vision toward which to collectively build.

The "problem-solving" approach to destructive environmental conflict emphasizes the capacity of facilitation to find solutions that generate mutually acceptable agreements, but almost always for the immediate economic benefit of adult humans, regardless of the effect of the agreement on children or the productive capacity of the environment. Facilitators using this approach often endeavor to influence and direct disputants toward agreement in general and even toward the specific terms of an agreement. This kind of facilitation of destructive environmental conflict fits well a Russian proverb: If men could foresee the future, they would still behave as they do now.

The "transformative" approach, in contrast to the problem-solving approach, emphasizes the capacity of facilitation for personal growth, which is embodied in the ability to accept risk. Transformative facilitators therefore concentrate on helping parties empower themselves to define the issues and come to agreement in their own terms and in their own time through a better understanding of one another's perspectives.

Transformative facilitators avoid the directiveness associated with the problem-solving approach. Equally important, transformative facilitators help parties recognize and capitalize on the opportunities for personal growth inherently present in conflict. This does not mean that satisfaction and fairness are unimportant; rather, it means that transformation of human moral awareness, conduct, and outcomes is even more important.

Speaking about transformation brings up the notion of change, or *conversion,* if you will, which means "turning" in both Hebrew and Greek. When you turn even slightly, you are going in a different direction, which leaves your old way of thinking farther and farther behind. Because change or turning in a new direction is frightening to people, the collective vision of a future toward which to build must be in place before a concept of sustainability is even relevant.

The second book in this series, *Sustainable Community Development: Principles and Concepts,* examines those principles and concepts that are necessarily a part of any sustainable community development. "Community," in the sense of sustainable development, focuses on the primacy and quality of relationships among people sharing a particular place and between people and their environment, particularly their immediate environment.

In the sense of sustainable community, "development" means personal and social transformation to a higher level of consciousness of cause and effect and a greater responsibility to be one another's keepers through all generations. Development in a community means asking how we can make the personal and collective transition to a high quality of life, one that is biologically, culturally, and economically sustainable. Development *does not* mean continual physical/economic growth, which is neither biologically nor economically possible without destroying the umbilicus between ecosystem and economy. (An ecosystem includes all living organisms interacting with their nonliving physical environment, considered as a unit.)

And "sustainability," in the sense of community development, is the act of one generation saving options by passing them to the next generation, which saves options by passing them to the next, and so on. Sustainability, in this sense, is a continual process, not some finite end point at which one arrives. As such, it will demand a shift in personal consciousness—from being self-centered to being other-centered, which brings us to this book, *Setting the Stage for Sustainability: A Citizen's Handbook,* the third in the series.

Having resolved the conflicts, working our way through the principles and concepts necessary to an understanding of sustainability, development, and community, we will now focus on preparing for sustainable community development, which means doing first things first.

Because we in the United States tend to ignore the notion of "first things first," this book is written with a U.S. audience in mind. We seldom take the time to figure out how best to do something before we leap into action, and when our actions do not produce the desired (or at times required) results, we repeat ourselves, but usually with frustration, seldom with greater wisdom.

Being in too much of a hurry to prepare properly before we act, our typical hurry-up mode is *Ready,* ***Fire****, Aim,* which means acting before we know where we want to go. Many a small rancher, for example, has gone bankrupt by following this hurry-up mode, not because he or she was not a good rancher, but because he or she found it easier to mend the fence rather than do the difficult work of ranch planning. "First things first" means slowing down and getting the sequence right: *Ready,* ***Aim****, Fire,* which means determining where we want to go *before* we act.

One of the authors (Chris) learned a valuable lesson about preparing to implement actions during the years he worked with the Shinto priests of Japan concerning the biological sustainability of their forests. The meetings he attended often seemed interminable, and he wondered if anything was getting done because the people (by his standard at least) seemed so indecisive.

They would discuss a point for an hour and seemingly accomplish nothing. But after they had examined the issue from its various points of view and prepared themselves, they came together and acted in far greater unison than we in the United States do. And in Chris's experience, they did not have to do things over and over again to get it right. They ensured the correctness of their actions, as best they could, *before* they acted, and then they planned with patience for the long term.

Such preparedness is a lesson we in the United States need to learn, especially where our social/environmental sustainability is at stake. The purpose of this book is to help with that lesson by examining some of the major things a community must address if it is to be sustainable. Once these things are addressed, each community can and must use its own ingenuity to figure out how to do what needs to be done because each community is unique in its desires and thus in the requirements those desires engender for social/environmental sustainability.

We, the authors, will be telling it as *we see it,* not necessarily as it is, which is all we can do because neither we nor anyone else *knows* how it is. Choosing a style of writing for this book has been relatively easy. We did our best to keep the book as conversational as possible so that we might engage the reader in a friendly discourse. Although our intended audience is city mayors, city planners, city council representatives, county commissioners, citizen activists, and university students of community planning and local government, we endeavored to make the text as widely readable as possible because communities belong to everyone.

On the other hand, choosing the level of sophistication and detail with which to write any given chapter has been difficult, in part because we do not know how disparate people's familiarity is with the different subjects covered in this book. Some chapters, such as Chapter 4 on communication, are particularly detailed and "cookbook-like" because, while almost everyone uses verbal communication, few people seem to know that most communication is nonverbal. Further, few people seem to pay much attention to their own skills in communication (especially those beyond the spoken word) and how well they can and do relate to others. Excellence in communication is becoming increasingly important as we in the United States become ever-more diverse and, therefore, conflict-prone and interdependent, both at home and abroad.

Other chapters, such as Chapter 11 on preparing to implement sustainable community, may seem to be written for more of a general audience. In this case, we are writing about principles and concepts that we find many people do not seem to recognize or understand, and if they do, they fail to understand their significance when dealing with social/environmental sustainability. We thus strove to assume a level of knowledge and detail that could be understood by a wide range of citizen activists.

We have purposefully *not* tried to write strictly a "cookbook" that can be followed by taking generic steps (as has been suggested by several people) because what works for one community may not work for another, and if it does work, it does so for different reasons. Further, were we to write this book in cookbook fashion, we would be enveloping you, the reader, in our own limits of vision, and that would be a gross disservice. Nevertheless, we have done our best to make this book as practical as possible, including an appendix of organizations and communities that can be contacted for information about their experiences with sustainable development.

Along these same lines, we have consciously written this book from an idealist's point of view, not because we think the majority of people are idealists but because the ideal is the only thing worth aiming at and therefore worth writing about. We believe the ideal to be within the grasp of human potentialities. Whether we humans are willing to struggle hard enough and long enough to reach that ideal is another issue, however, and not part of this book.

Because the second book builds on the first and this third book builds on the first two, there are some ideas carried forward into this third book. Although such repetition is necessary because each book is designed to stand on its own, it is minimized as much as possible.

Chapter 1 of this book is written by Chris (with Russ and Kevin's permission) to carry forward the observations of change within a community, which was begun in the second book. The purpose is twofold: (1) to unite the second and third books and (2) to set the stage for this book.

As you read this book on sustainable community development, remember the words of botanist George Washington Carver: "99 percent of the failures come from people who have the habit of making excuses."

ACKNOWLEDGMENTS

First, I, Chris, extend my heartfelt appreciation to Russ Beaton and Kevin Smith, my two co-authors, for helping to make the experience of writing this book educational, rewarding, and fun.

Second, the following people, in alphabetical order, were kind enough to review the manuscript and make substantive improvements: Joe Bowersox (Assistant Professor of Politics, Willamette University, Salem, Oregon), Mike Omeg (undergraduate student in economics and environmental science, Willamette University, Salem, Oregon), Todd Sexton (undergraduate student in economics and environmental science, Willamette University, Salem, Oregon), Jane Silberstein (Planning and Community Development, Washburn, Wisconsin), Laura Tilley (Adjunct Professor, Portland State University, Portland, Oregon), and Nathan Young (undergraduate student in economics and environmental science, Willamette University, Salem, Oregon). Thank you.

The three students, Mike, Todd, and Nathan, but especially Nathan, gave sensitive and integrated counsel beyond their years as they reviewed the manuscript.

We are also grateful to Sandy Pearlman for the distinct pleasure of working with her once again as she helped our writing stretch from good toward excellent.

To my wife, Zane, I, Chris, again extend my profound appreciation for her patience not only as I worked on this book but also for the hours she spent proofreading the pages, a critical task toward the excellence of any book.

To my wife, Delana, I, Russ, extend my special "thank you," because, as is usually the case with such efforts as this book, Delana bears the initial costs.

AUTHORS

Chris Maser spent over 20 years as a research scientist in natural history and ecology in forest, shrub steppe, subarctic, desert, and coastal settings. Trained primarily as a vertebrate zoologist, he was a research mammalogist in Nubia, Egypt (1963–64) with the Yale University Peabody Museum Prehistoric Expedition and was a research mammalogist in Nepal (1966–67) for the U.S. Naval Medical Research Unit #3 based in Cairo, Egypt, where he participated in a study of tick-borne diseases. He conducted a three-year (1970–73) ecological survey of the Oregon Coast for the University of Puget Sound, Tacoma, Washington. He was a research ecologist with the U.S. Department of the Interior, Bureau of Land Management, for 12 years (1975–87), the last 8 studying old-growth forests in western Oregon, and a landscape ecologist with the Environmental Protection Agency for a year (1990–91).

Today he is an independent author as well as an international lecturer and a facilitator in resolving environmental disputes, vision statements, and sustainable community development. He is also an international consultant in forest ecology and sustainable forestry practices.

He has written over 250 publications, including the following books: *Forest Primeval: The Natural History of an Ancient Forest* (1989, listed in the *School Library Journal* as best science and technical book of 1989), *Global Imperative: Harmonizing Culture and Nature* (1992), *Sustainable Forestry: Philosophy, Science, and Economics* (1994), *From the Forest to the Sea: The Ecology of Wood in Streams, Rivers, Estuaries, and Oceans* (1994, with James R. Sedell), *Resolving Environmental Conflict: Towards Sustainable Community Development* (1996), and *Sustainable Community Development: Principles and Concepts* (1997). Although he has worked in Canada, Egypt, France, Germany, Japan, Malaysia, Nepal, Slovakia, and Switzerland, he calls Corvallis, Oregon, home.

Russ Beaton is Professor of Economics at Willamette University in Salem, Oregon. He co-authored Oregon's nationally recognized legislation on land-use planning and has long been involved, both as researcher and citizen, in affairs dealing with energy, the environment, land use, and local and regional economic issues. He has authored studies on timber, agriculture, and urban growth. Among other topics typical of a liberal arts college faculty member, he teaches energy economics, environmental economics, and a course titled "Regional Economics and the Economy of Oregon." Russ lives in Salem, Oregon, and intends to spend the rest of his career involved with issues surrounding the notion of sustainability.

Kevin Smith, Chief of Staff for former Oregon Governor Barbara Roberts, is currently the Intergovernmental Affairs Manager for the Oregon Economic Development Department, where he works with community-based collaborative stewardship organizations in helping create new policies that support sustainable development. He manages interagency, interstate, and congressional programs relating to local, state, and federal issues, including sustainable development and economic adjustment partnerships for natural-resource-based rural communities. He is a member of both Governor John Kitzhaber's Healthy Environment Policy Team and the Coastal Salmon Restoration Initiative and also serves on the Western Regional Council of the President's Council on Sustainable Development. He attended the State and Local Government program at Harvard's John F. Kennedy School of Government and is a graduate of Portland State University. Kevin lives in Washington, D.C.

RETURNING HOME 1

In Chapter 1 of the book *Sustainable Community Development: Principles and Concepts,* I (Chris) chronicled a few of the events, beginning in the early 1940s, that changed my hometown. I watched its transformation from a friendly, safe community (where we did not lock the doors to either our homes or automobiles; had milk, cream, butter, cottage cheese, and eggs delivered each morning to our front doors; did all of our shopping downtown on foot; knew the shopkeepers; and visited with people in the street) to one where people are murdered; women are attacked and raped, even in broad daylight; urban sprawl has gobbled up much of the surrounding farmland and some forestland; small, diversified family farms have given way to large, monocultural, corporate-style farms; and one must now drive hither and yon among shopping malls merely to fulfill normal living requirements.

Having been raised in a small community which in the days of my youth would have been called "sleepy," I could not, as a young man, wait to leave it and explore the world, which I did for almost two decades, both in the United States and abroad. Then, a few years ago, I came back to my hometown to live. I was not, however, prepared for how much it had changed.

It has grown in size and now sprawls over the countryside into the irreplaceable agricultural lands. The congestion of traffic is incredible and rapidly growing worse. Reported child abuse is on the rise, as is the resulting violence in the streets.

In short, my hometown has lost its feeling of friendly trust and sleepy innocence. It is now a town of guarded people, where I cannot pick up a child unknown to me and help him or her get a drink from a public water fountain without putting myself in legal jeopardy.

A recent article in the local newspaper was written by a woman who lives in town and felt threatened—and angry because of it—when she was approached in a lonely place by a strange man. It made no difference that the man was of friendly demeanor and innocent intent. The fact that the woman had gone alone to find solitude in Nature is not the issue, because she has as much right to find solitude in Nature as a man does.

I understand her feelings of fear and the anger they cause. My own wife is also afraid to venture alone into Nature in the vicinity of my hometown, especially at night to enjoy the stars, whereas I, even as a child, was free to wander at will without fear. It makes her angry too; after all, she has as much right to a life among other people without fear as I do.

"Ah yes," you might say, "but is not her fear a choice and all in her mind?"

Perhaps this is true in a psychological sense, but the reality of such fear in today's society is not so easily dismissed, which gives me pause to wonder: What has happened to my hometown? What has happened to its people? It is no longer safe, especially for women and children, to travel alone in Nature because of the violence in our society—and my hometown.

When I came back to my hometown to live, I realized that I had matured enough to want to give something to the place in which I chose to reside, so I accepted a three-year appointment on the Environmental Advisory Committee in my county. The committee's charge was to advise the county commissioners on environmental issues.

Even after having lived in the red shadow of communism and under the black threat of a dictator's fist, and having spent more than a decade as a federal research scientist in the U.S. Department of the Interior, Bureau of Land Management, I still naïvely thought *local* government (especially that of my own county and hometown) to be more sincerely democratic and to perform with other-centered honesty, forthrightness, farsightedness, enthusiasm, service, and authenticity. I thought the representatives of local government sincerely cared about the generations of the future. After all, it is the physical, emotional, and spiritual welfare of their own children and grandchildren at stake. What I found instead made me sick at heart.

During my tenure on the committee, for example, the need for a county/city vision for the next century was repeatedly brought up. Four years later, there is still no action. How can we plan wisely and monitor the results of our planning intelligently if we do not know where we want to go, much less where we are going accidentally?

The urgent necessity for both the county and city to study the long-term (at least the next century's) potential supply of sustainable, potable water by comparing available, healthy water catchments with the county/city requirements (in the face of rapidly disappearing open space and increasing human population) was brought up repeatedly. To date, the imperative of planning for sustainable water catchments has been omitted from meetings I have attended on open space (with the exception of one meeting in which my discussion group specifically brought it up). It has also been absent in any of the meetings on proposed housing developments that I have attended or other land-use plans that I have seen. And I do not remember its having been treated in the county's comprehensive plan.

In fact, while I still served on the Environmental Advisory Committee, the city announced plans to increase the capacity of its water treatment plant by 5,000 homes. I recommended that only 3,000 homes be built and that a 2,000-home margin of safety in the supply of water be held in reserve for emergencies. The response was "We can't do that. We can't limit growth. We'll figure something out when we reach the critical capacity and find we need more water." The current government thus chose—once again—to pass the bill to the future rather than take a "pay as you go" responsibility for itself.

Who is willing to be responsible and when?

A subcommittee of the Environmental Advisory Committee also wrote a two-paragraph preamble to the county's comprehensive land-use plan, which in reality was merely a legally forced response to a mandate imposed by the state—and read as such. The preamble was intended simply to put the coerced plan into a more positive perspective, but it contained two words—*accountability* and *sustainability*—that so disturbed county officials that adoption of the preamble was paralyzed for at least the remaining year and a half of my term. The reason given was the need for more study.

Much of the risk-avoidance behavior in today's government agencies dealing with land use is not only about preserving the bureaucratic system but also about being the bull's-eye in the cross hairs of the rifles of both industrialists and environmentalists, each of which has a finger on its respective trigger. Because such behavior elicits a "duck and run for cover" posture, not a single thing of merit was accomplished during my tenure on the committee, other than confusing motion with accomplishment, a tragically common way people have of avoiding accountability.

I think of this as "shadow government" because there is nothing and no one to get a hold of. Certainly, committees are established that look good and are politely listened to, but for the most part they have no real authority and so stand as toothless dogs guarding the future. It is therefore necessary to rely on transformative facilitation to end the polarization and scorched-earth litigious nature of the opposing sides.

I could go on at length (as you probably also could from your own experience), but I will defer momentarily to George Bernard Shaw, who wrote: "You see things; and say 'why?' But I dream things that never were: and I say 'why not?'" Although I clearly am no George Bernard Shaw, I dream of sustainable community development and ask, "Why not and how?" The government officials see things that are and say, "We can't; it's not ours to do. Maybe later, not now."

I saw a poem some years ago for which I found no author, but it is relevant here:

One ship goes East,
    another West
By the self-same winds
    that blow

'Tis the set of the sails
    and not the gales
That determines the
    way they go

Like ships at sea are the
    ways of fate
As we voyage along
    through life

'Tis the set of the soul
    that decides the goal
And not the calm
    or strife

I now see that local government can be just as dysfunctional as the federal government, despite a supposedly increased ability to relate to those whom it serves. But then, government is only a collection of people and is, of course, as functional or dysfunctional as they are. (See *Resolving Environmental Conflict*[1] for a discussion of dysfunction at the family level and *Sustainable Forestry*[2] for a discussion of dysfunction at the agency/government level.)

Dysfunction in a government comes visibly to the fore when the production of a product (such as a housing development on prime agricultural land coveted by a politically strong individual or special interest group) becomes more important than the process of human relationships that fulfill the campaign promises made to the people as a whole. The notion of "accountability" has thus produced a "product fetish" among government officials because they feel a requirement to provide tangible proof of their worth to those people they think can help them retain their jobs.

Dysfunction becomes particularly apparent when being reelected, reappointed, or otherwise keeping one's job takes precedence over the necessities of the people one is sworn to serve. Another symptom occurs when maneuvers of keeping one's job are more important than the dignity and well-being of future generations or even one's personal integrity. What happens?

Rather than recognizing that the people at large are both the bosses and the customers of the government and treating them accordingly, individuals in the government often bow to monied interests (be they corporate or otherwise) through political pressures and serve those who either match their own interests and/or can affect the future of their jobs. Again, the way our political system is practiced tends to serve the private power base, not the people at large. However, careful rhetoric and packaging define this behavior as acting "in the public interest."

The theory of our democratic government is that people act collectively to govern themselves. But when a government becomes dysfunctional and thus isolated from the people at large, it feels threatened by changing social values and takes on a life of its own for the preservation of the status quo. At such a time, it sees the very public it is meant to serve as its enemy.

When such dysfunction is pointed out in news stories and court cases, the "views and policies" of the government become the backbone of its homeostatic defense, a defense that is usually upheld by the courts if legal procedure has been followed, regardless of substantive testimony and social/environmental consequences. I know this having served more than once as an expert witness, where I repeatedly observed this type of outcome.

Homeostasis refers to a system that controls the behavior of a group of people (usually a family) through an enforced set of rules that determine the acceptable interactions of its members in organized, established patterns. The rules are a prescription aimed at maintaining a dynamic equilibrium within the group by defining each member's relationship to the whole. Homeostasis thus provides a mechanism through which the group's dynamics protect its secret of dysfunction while making it look good outwardly by keeping its name pure.

Once a government has become dysfunctional, it begins to perform institutionalized rituals to ensure its survival and hide its dysfunction. It can, however, be dysfunctional long before it is so perceived.

In today's dysfunctional government, therefore, those ideas contrary to the established view are termed "heresy," and those that implicitly support the established view are termed "policy." Policy is thus synonymous with homeostasis, which determines what shall and shall not occur within and outside of the government.

Homeostasis also becomes a mechanism through which a government's inner workings are kept hidden from "outsiders," while appearing to operate totally openly by obeying such laws as those pertaining to open meetings. This is an important concept because a government, be it local or national, is only a collection of people who ostensibly have come together to serve the public—not to make the public serve them.

But for a blindly "loyal," unthinking, unquestioning government employee, or one afraid of losing his or her job, to be truly a servant of the public is an intolerable position. It is intolerable because the homeostasis of the government (and therefore the security of the person's known position within it) is always in jeopardy from a *thinking* "outsider" who has the persistent audacity to ask unfavorable questions.

Such concern is to some extent reasonable in that to function, even within its original charge, a government must work as a team, and one uncaring maverick can disrupt the whole effort. On the other hand, homeostasis within a dysfunctional government can hide corruption because it is a conspiracy of silence—or "cover-up," if you will.

To safeguard its "good" name, a dysfunctional government becomes a self-serving machine that often defines more and more narrowly and more and more rigidly the job descriptions of its employees and thereby controls them. Like loss of flexibility in an ecosystem groomed by humans to produce specialized products (such as a hundred acres of prairie being converted into a wheat field), the rules of homeostasis become ever-more inflexible and exacting. In this sense, the description of an employee's job directs the employee to do something, but the dysfunction of the government simultaneously requires that the employee must be prevented from carrying out the assignment in a 100 percent professional, questioning manner if the government's deception of the public is to be contained.

Keeping the dysfunction under wraps is a test of blind loyalty to the government and its value system, which is seen as a *good* thing, even though it mimics the tyrannical inequities our democratic government was originally designed to correct. Failure to pass the test usually has severe consequences.

Consider, for example, the following story about the military in which a man did exactly as he was supposed to do within his job description and also did exactly as he was supposed to do as an ethical, professional employee:

> An Army officer who directed an inventory that uncovered improperly diverted explosives at Fort Lewis [Washington] is trying to win another command, after he resigned his previous post because of conflicts over the incident.
>
> Flick said his role in identifying the mismarked and diverted munitions has cost him his career because he broke a bond of trust by going over his commanding officer [the immediate source of dysfunction] to point out the irregularities.
>
> "They feel that I broke the bond of trust, even though the part I did was lawful [within his professional job description and Army regulations] and what they did was unlawful...."[3]

Few people have the courage to stand up and speak out for what they know is right in the first place. And when they do, they are seldom rewarded for their honesty and loyalty to their ideals as professionals, because preventing these kinds of "leaks" is what homeostasis, as the term has been used here, is all about. In some cases the homeostasis fails and change takes place, but homeostasis unfortunately succeeds in most cases, and an untold number of truly loyal people die a slow, unknown "professional death" for their integrity and courage both as human beings and true servants of the public.

A key to homeostasis is the job description. Although a job description is necessary and seems harmless in itself, it can be used either to relate the work of professionals to one another and among disciplines or to isolate professionals and disciplines from one another. The danger lies in the isolation of individuals and disciplines by increasing professional specialization as manifested through the purposeful narrowing of the interpretation of job descriptions and through the careful, rigid, and absolute control of those descriptions.

The use of job descriptions as an isolating mechanism is one of the ways a government protects its members from "unfavorable" information. The institutionalized, internal policies of the government shape the perceptions and beliefs of uncritical employees in ways that protect the government's dysfunction, even when catastrophic outcomes are involved. In fact, these internal policies are vigorously used to justify the necessity of narrowly trained specialists and experts.

A distortion of information is not limited to willful deceit on the part of an individual who is perceived as loyal to the government. Even honesty within a dysfunctional government is insufficient to prevent the widespread distortion of information. The weakness lies within the job description itself and with an individual's acceptance of an assignment without thinking about or questioning the consequences of completing it.

Although this hardly sounds untrustworthy, much less dangerous, it is just this "functionary" behavior that allows systematic, homeostatic distortions to occur. This is why decisions are often difficult to deal with in governments, because one is seldom sure who makes them. Almost everyone seems to be afraid of taking a risk; therefore, if more than one person is responsible for a decision, especially a miscalculation, no one will be at fault. (Having said this, however, I point out that efforts are under way to honor and promote "risk taking" within government. I speak of Vice President Gore's National Performance Review and its challenge to managers to reward innovation that requires risk.)

One thus becomes a functionary by limiting one's inquiries only to questions of how best to accomplish an immediate assignment while allowing the government to shape one's perceptions. The fault, therefore, lies neither in the job description nor in the assignment. The fault lies in not accepting personal responsibility for the outcome of the assignment, which de facto is accepting current homeostatic rules for one's personal behavior. This is the seed of dysfunction, as well as the birth of the machine which steals personal identity, individuality, personal integrity, and ultimately human dignity.

The corporate/political machine tells us that the economic inequality that we are now experiencing is a regrettable but necessary consequence of a free market economic system. The machine says that environmental degradation (although unfortunate) is an inevitable, even necessary, trade-off if economic prosperity is to be maintained. Taken together, we are being told through such statements that there is little or no role for a civil society, let alone a civil community, in shaping and directing an economy that is of the people, by the people, and thus good for the people and their environment, present and future.

And the pundits' views are authoritatively stated with considerable disdain for our constitutional commitment to a free, participatory democracy. Many officials in local, state, and national governments either concur with the corporate/political pundits, as evidenced by their inaction, or they are in positions of leadership with myopic minds and no ability to lead.

In short, we cannot maintain the essence of the American ideals, as envisioned so long ago, while simultaneously degrading our environment, undermining our democracy, and destroying our families and communities all in the name of an increasingly fragile, short-term, elitist economic system. The logic, which assumes that it is impossible for our economic system to embrace values of equality, social/environmental sustainability, and inclusive participatory democracy, is not only flawed but also dictatorial and thus decidedly antidemocratic.

Therefore, simply taking orders without thinking about or questioning them may be personally safe but is environmentally and socially risky; I know this to be true from personal experience. On the other hand, it is often personally risky—if you want to keep your job—to question orders (again, I know from personal experience), but questioning orders is both environmentally and socially responsible and is exactly what a good public servant should be hired to do. Besides, "responsibility educates," as Wendell Phillips, an American abolitionist, observed.

Until transformative facilitation replaces polarization, most public servants (and consultants) will continue to be told, albeit subtly, what level of professionalism to practice in order to stay employed. People therefore trade their dignity and professional ethics for job security.

Unfortunately, these frightened employees are often judged harshly for being functionaries, but this judgment is usually rendered by people outside of the "system" who do not have the benefit of understanding how it works. I, however, understand the inner workings of a dysfunctional government because I had to resign from the Bureau of Land Management almost a decade ago to protect my integrity, which—like virginity—once forsaken is forever lost to me personally.

Nevertheless, I expected better from local government. But I stand corrected and hark back to the words of American writer Carl Sandburg: "If America forgets where she came from (the "self-evident" truths upon which the American experiment is based), if the people lose sight of what brought them along, if she listens to the deniers and mockers, then will begin the rot and dissolution."

Now I turn a deaf ear when I hear local people rail at our federal government, saying that local government does a better job of managing the land in the present for the generations of the future. I have seen no proof of that claim at any level of local or state government, where commercial fishing, livestock grazing, tree farming, or mining is concerned. But I have seen individual ranchers and woodlot owners do outstanding jobs as trustees of the land for the next generation while making a living for themselves. Unfortunately, I know of none in local government.

Be that as it may, it is exactly because of these outstanding individuals that I have hope for society. Through conscious personal growth, we can all become outstanding individuals who collectively choose to create communities to be catalysts and to instigate beneficial change within our own government and our immediate environment. Such change will come about by developing private/public partnerships, where community members and government employees sit down together and ask: "If not us, then who will transform our government while dreaming things that never were and asking 'why not?'"

This book is offered as a step in creating such catalytic communities of outstanding individuals, beginning with a renewed look at the meaning of community.

2 THE CONCEPT OF COMMUNITY

Community, as English historian Arnold Toynbee said of civilization, "is a movement and not a condition, a voyage and not a harbor." But community is a deliberately different word than civilization or even society. Although community may refer to neighborhoods or workplaces, to be meaningful it must imply membership in a human-scale collective, where people encounter one another face to face.

Community is thus a group of people with similar interests living under and exerting some influence over the same government in a shared locality. Because they have a common attachment to their place of residence, where they have some degree of local autonomy, they form the resident community.

People in such a community share social interactions with one another and organizations beyond government and through such participation are able to satisfy the full range of their daily requirements within the local area. The community also interacts with the larger society, both in creating change and in reacting to it. Finally, the community as a whole interacts with the local environment, molding the landscape within which it rests and is in turn molded by it. In this sense, community is about the oneness of the whole and the wholeness of the one.

TRUE COMMUNITY IS FOUNDED ON A SENSE OF PLACE, HISTORY, AND TRUST

Community is rooted in a sense of place through which the people are in a reciprocal relationship with their landscape. As such, a community is not simply a static place within a static landscape, but rather is a lively, ever-changing, interactive, interdependent system of relationships. Because a community is a self-organizing system, it does not simply incorporate information, but changes its environment as well. Thus, as the community in its living alters the landscape, so the landscape in reaction alters the community.

Reciprocity is the self-reinforcing feedback loop that either extends sustainability to or withholds it from a community and its landscape. We therefore create trouble for ourselves in a community when we confuse order with control. Although freedom and order are partners in generating a viable, well-ordered, autonomous community, a community is nevertheless an open system that uses continual change to avoid deterioration.

A community also has a history, which must be passed from one generation to the next if the community is to know itself throughout the passage of time. History is a reflection of how we see ourselves and thus goes to the very root by which we give value to things. Our vision of the past is shaped by, and in turn shapes, our understanding of the present—those complex and comprehensive images we carry in our heads and by which we decide what is true or false.[4]

When the continuity of a community's history is disrupted, the community suffers an extinction of identity and begins to view its landscape not as an inseparable extension of itself but rather as a separate commodity to be exploited for immediate financial gain. When this happens, community is destroyed from within because trust is withdrawn in the face of growing economic competition.

It seems clear, therefore, that true community literally cannot extend itself beyond local place and history. Community, says Wendell Berry, "is an idea that can extend itself beyond the local, but it only does so metaphorically. The idea of a national or global community is meaningless apart from the realization of local communities."[5]

For a true community to be founded in the first place and to be healthy and sustainable, it must rest on the bedrock of mutual trust among its members.

> ...a community does not come together by covenant, by a conscientious granting of trust. It exists by proximity, by neighborhood; it knows face to face, and it trusts as it knows. It learns, in the course of time and experience, what and who can be trusted. It knows that some of its members are untrustworthy, and it can be tolerant, because to know in this matter is to be

> safe. A community member can be trusted to be untrustworthy and so can be included. But if a community withholds trust, it withholds membership. If it cannot trust, it cannot exist.[5]

"Trust," according to the *American Heritage Dictionary,* "is firm reliance on the integrity, ability, or character of a person or thing; confident belief; faith." But trust cannot really be defined because it is based on faith that a particular person is "trustworthy" or faithful to his or her word. Trust can only be lived in one's motives, thoughts, attitude, and behavior.

Trust versus mistrust is the psychosocial crisis in the first of Erik Erikson's eight stages of human development.[6] Trust versus mistrust is the dominant struggle from ages 0 to 1. Erikson assigned hope as the virtue of this stage in which the mother–baby relation lays the foundation for trust in others and in oneself. But as everything has within itself the seed of its opposite, this stage also presents the challenge of mistrust in others and a lack of confidence in oneself.

Hope, as the virtue of trust, is the enduring belief that one can achieve one's necessities and wants. Trust in human relationships is thus the bedrock of community and its sustainable development.

If trust is not developed, none of Erikson's other stages of development can take place: *autonomy* versus shame and doubt, *initiative* versus guilt, *industry* versus inferiority, *identity* versus identity confusion, *intimacy* (relationship) versus isolation, *generativity* versus stagnation, and *integrity* versus despair—all of which are part and parcel of community sustainability. Community is therefore the melding of how people in different developmental stages relate to both themselves as individuals within a community and with others as a community.

In sum, community is relationship, and meaningful relationship is the foundation of a healthy, sustainable community. In this connection, Ralph Waldo Emerson penned the following: "It is one of the most beautiful compensations of this life that no man can sincerely try to help another without helping himself." William James said it thusly: "Wherever your are, it is your own friends who make your world."

As such, a resident community serves five purposes:[7] (1) social participation—where and how people interact with one another to create the relationships necessary for a feeling of self-worth, safety, and shared values; (2) mutual aid—services and support offered in times of individual or familial need; (3) economic production, distribution, and consumption—jobs, import and export of products, as well as the availability of such commodities as food and clothing in the local area; (4) socialization—educating people about cultural values and acceptable norms; and (5) social control—the means for maintaining those cultural values and acceptable norms.

Community also reminds one that the scale of effective organization and action has always been small local groups. As anthropologist Margaret Mead says: "Never doubt that a small group of thoughtful, committed citizens can change the world; indeed it is the only thing that ever has."

It is therefore logical that community not only is a way of valuing the independent voluntary or nonprofit organization but also relies for its expression on such institutions as neighborhood schools, family centers, and volunteer organizations. Further, creating sustainable communities strengthens one's fidelity to a sense of place and is the best possible immigration policy because it raises the value of staying home. These things top-down government cannot fulfill.

With the current disintegration of family and local community in American life, it is unlikely that most people in this country really have an intimate sense of trust and belonging. We have largely lost our sense of connection to and with community that once impressed the French political figure and traveler Alexis de Tocqueville to the point that he wrote in the 1830s:

> Americans of all ages, all conditions, and all dispositions constantly form associations...religious, moral, serious, futile, general or restricted, enormous or diminutive. The Americans make associations to give entertainments, to found seminaries, to build inns, to construct churches, to diffuse books, to send missionaries to the antipodes; in this manner they found hospitals, prisons and schools.[8]

He went on to argue that is was no accident that "the most democratic country on the face of the earth is that in which men have, in our time, carried to the highest perfection the art of pursuing in common the object of their common desire." Why then the progressive disintegration of trust?

Consider that in 1966 only 30 percent of the people surveyed said they did not trust the government in Washington, D.C., some of the time or all of the time, and in 1992 75 percent of the people surveyed responded in the negative.[8] What has happened to the most democratic country on Earth? Why have we lost our sense of community? There are at least two possibilities.

One reason for this loss of community may be our lopsided expansionist economic world view in which material possessions and the incessant push for continual economic growth take the place of spirituality, as once manifested in quality relationships and mutual caring. The economic world view translates into both adults in many households having to work at paying jobs outside the home just to make ends meet, which raises the question of who is left at home to act as a parent and forge community ties when both adults are too busy. If, however, human society and its environment are ever to become sustainable, it is necessary to rediscover or recreate our sense of local community in order to balance the material with the spiritual, the piece within the whole.

The other reason is summed up by Abraham Maslow: "We [as human beings] fear our highest possibilities (as well as our lowest ones). We are generally afraid to become that which we can glimpse in our most perfect moments, under the most perfect conditions, under conditions of greatest courage. We enjoy and even thrill to the god-like possibilities we see in ourselves in such peak moments. And yet we simultaneously shiver with weakness, awe, and fear before these same possibilities."[9] Is it, as Maslow says, our fear of our own greatness and success that is the inner enemy made manifest in the moral decay that is consuming communities in this country?

LOCAL COMMUNITY UNDER STRESS

Although the last two centuries may have nurtured such institutions as political freedom and the rights of private property, they have done little for the quality of relationship—the trust—that holds traditional (resident) American communities together. The last two centuries have done even less to nurture the concept, let alone the reality, of multiracial communities. But this softer value of trust is the social capital that enables people to work together and commit to common causes. As such, relationships of high quality and integrity are absolutely critical to the success of a community in translating its cultural identity into a shared vision of the future toward which to build.

For a community to fulfill its vision, it must be grounded in personal ethics, which are translated into social ethics. This puts the responsibility for one's own conscience and behavior where it rightfully belongs—squarely on one's own shoulders. With a strong sense of personal and social ethics, communities will be spared wasting time and money policing socially unacceptable behavior, which ultimately leads to the destruction of both a community and its landscape. With a strong sense of personal and social ethics based on a solid foundation of spiritual consciousness, neither the environment nor future generations will be the dumping ground for personal and social irresponsibility.

While for some community may simply be a useful new concept to wrap around old ideas and institutions, for others it will be a new set of ideas, a new frame of reference about how and why people relate to one another and to the wider world. Its value lies in creating a bridge between people's core values and principles for action and governance, which will help shift perceptions about what politics and government are really for, such as balancing growth in the population of a community with the biologically sustainable capacity of its landscape.

There is in each community an upper limit to population beyond which the overall quality of life becomes unalterably diminished and the immediate landscape irretrievably damaged with respect to human values. If an upper limit to population size is selected, a certain percent of the population would reply that the community is already too large. Another segment of the community would respond that it is big enough right now. Still another portion would assert that if things were done better, such as all phases of careful strategic planning, a much larger population could be accommodated. But at some point, no matter how well growth is "managed," people realize that the upper limits of a population do count.[10]

Once a certain size, a critical mass, is reached, there is no reclaiming the more comfortable scale of a community with its ambiance. Beyond some point, the indescribable sum total of population, traffic, human activities, commotion, noise, the clarity of stars at night, the quality of air and water, demand on public services, quality of personal relationships, and the ability to put it all in the context of a place of *quality* called "home" is irretrievably lost.[10]

Why should 10 to 15 percent of the people (those who will never accept limits to a community's size because of vested interests in continual growth) be able to determine the long-range future of a community's population? Urban Growth Boundaries are designed to limit the size of a community within the context of its landscape as long as development meets certain criteria. But those with vested interests in continual growth would push for increasing the size of the Urban Growth Boundary once filled to capacity, which negates the whole concept of containing population growth.[10]

So the question becomes one of how a community can be given the legal right to limit the growth of its population if it so chooses. The decision, once made, needs to be written into the city charter and used to provide a context for all documents and any further planning.[10]

Amending the city charter is important because there is an increasingly common yearning for a more defined and authentically lived set of ethical values (trust) with which to rekindle meaning and purpose in life and politics. The language of community is one way of reconnecting people to a set of shared values and principles with which to embrace daily uncertainties.

Shared ethics must be nurtured as one of the most valuable assets in making human communities work. For a community to work, it must nurture human-scale structural systems within which people can feel safe and at home in a particular place to which they feel a measure of fidelity. And it is precisely this sense of safety in and fidelity to a particular place that is being called into question as the face of community is being redefined in a more worldly context.

SHADES OF COMMUNITY: A LESSON FROM BIRDS

"What," you might ask, "is meant by a measure of fidelity?" Here it is instructive to consider communities of birds in a given area as ornithologists think of them. First, there is the resident community, which is that group of birds inhabiting the area to which they have a strong sense of fidelity all year. In order to stay throughout the year, year after year, they must be able to meet all of their ongoing requirements for food, shelter, water, and space. These requirements become most acutely focused during the time of nesting, when young are reared, and during harsh winter weather.

Then there are the summer visitors, which overwinter in the southern latitudes and fly north to rear their young. They arrive in time to build their nests, and in so doing must fit in with the year-long residents without competing severely for food, shelter, water, or space, especially space for nesting. If competition were too severe, the resident community would decline and perhaps perish through overexploitation of the habitat by summer visitors, which have no lasting commitment to a particular habitat.

There are also winter visitors, which spend the summer in northern latitudes, where they rear their young, and fly south in the autumn to overwinter in the same area as the year-long residents, but after the summer visitors have left. They too must fit in with the year-long residents without severely competing with them for food, water, shelter, and space during times of harsh weather and periodic scarcities of food. Here, too, the resident community would decline and perhaps perish if overexploitation of the habitat through competition were too severe. And like the summer visitors, the winter visitors are not committed to a particular habitat but use the best of two different habitats (summer and winter).

On top of all this are the migrants, which come through in spring and autumn on their way to and from their summer nesting grounds and winter feeding grounds. They pause just long enough to rest and replenish their dwindling reserves of body fat by using local resources of food, water, shelter, and space, to which they have only a passing fidelity necessary to sustain them on their long journey.

The crux of the issue is the carrying capacity of the habitat for the year-long resident community. If the resources of food, water, shelter, and space are sufficient to accommodate the year-long resident community as well as the seasonal visitors and migrants, then all is well. If not, then each bird in addition to the year-long residents in effect causes the area of land and its resources to shrink per resident bird. This, in turn, stimulates competition, which under circumstances of plenty would not exist. If, however, such competition causes the habitat to be overused and decline in quality, the ones who suffer the most are the year-long residents for whom the habitat is their sole means of livelihood.

Here we might anticipate your question concerning what a resident bird community has to do with a resident human community. It has to do with the statement previously made by Wendell Berry, that a true community can extend itself beyond the local, but *only* if it does so *metaphorically*.[5] This means that if the resident community is rendered nonsustainable by outside influences, such as people from other areas overharvesting local crops of mushrooms or large absentee corporations clear-cutting forests to the detriment of local water catchments, then the trust embodied in the continuity of a community's history is shattered, as is the self-reinforcing feedback loop of mutual well-being between the land and the people.

Another, more subtle way outside influence can destroy community is transients in its population. In one small town in Idaho, where we asked people how they felt about the fairly large number of employees of the U.S. Forest Service living in their community, they replied that they tried *not* to get to know them.

When asked if they avoided getting to know the folks from the Forest Service because they were transients who felt no sense of place within the community, the answer was only partly in the affirmative. They said it was mainly just too painful to become friends with and learn to trust Forest Service employees only to have them leave in two or three years. That kind of continual loss was too much like perpetual grieving for the death of friends and was more than the community could abide.

When a community loses (for whatever reason) the cohesive glue of trust embedded in its fundamental values, it loses its identity and is set adrift on the ever-increasing sea of visionless competition both within and without, where "growth or die" becomes the economic motto driving the cultural system. Such visionless competition inevitably rings the death knell of community.

THE EXISTENCE OF COMMUNITY DEPENDS ON HOW WE TREAT ONE ANOTHER

To protect a resident community's sustainability within its landscape, the community's requirements must be met before other considerations are taken into account; if this does not happen, no other endeavor will be sustainable. Let's consider, for example, the First Americans.

First Americans

Prior to the invasion of foreigners from Europe, the First Americans had unlimited natural resources per capita on a long-term basis, although such resources as food may have been limited seasonally. (Today, however, local communities face increasingly permanent limitations on natural resources, both renewable and nonrenewable, in addition to which seasonal limitations must also be taken into account.)

When the Europeans arrived and began competing for those same resources, the inevitable outcome was not readily apparent. But as the numbers of Europeans continually increased, through both local births and rapid immigration, the First Americans were increasingly pushed out of the way by the seemingly limitless numbers of Europeans, each of whom demanded his or her "fair share" of the available resources.

Moreover, the First Americans shared the land, whereas the Europeans took forcible ownership thereof to the exclusion of a whole indigenous cultural myth. The Europeans superimposed their mythology on that of the First Americans and consciously set about destroying not only the culture of the First Americans but also the mythology upon which it was based. With the demise of their resources and culture, the First Americans lost their sense of place, hence their sense of mythology, hence their sense of identity, hence their sense of community, and finally their cultural soul.

"Corporate" Towns

Many of today's local communities are in a similar type of jeopardy as were the First Americans because they are little more than the economic colonies of large national and international corporations. The corporations—whose fidelity is to the profit margin, not a sense of people, community, or place—increasingly siphon off as much of the local capital as possible and give as little in return as possible. (This is how the Europeans treated the First Americans.)

When the corporations withdraw their presence, because the resources on which they count become depleted or markets fail, communities, which were built around the corporations, are left to fend for themselves. This often means that they must use all available natural resources in their local landscapes if they are to diversify enough to survive.

Outside Pressures

Today, local communities, like resident bird communities, are facing increasing outside pressures from people who move seasonally into their local landscapes to harvest renewable natural resources, which the local people themselves often need to survive as a community. When the harvest is over, the seasonal visitors leave. The issue, therefore, is no longer job stability but rather community sustainability. And because the people coming into the area are often from other countries and nationalities, there is a clash of mythologies and a corresponding lack of communication, as often happens when people are forced together through perceived necessity.

In some cases, people moving into an area for the seasonal harvest of its resources, like the birds, may be able to fit in without overharvesting. There is a caveat to this statement, however: There must be a firm limit to the number of seasonal gleaners and the quantity they harvest, which unequivocally takes into account the requirements of the local residents and the sustainable productive capacity of the landscape.

In other cases, where the necessities of the local year-long residents and the sustainable productive capacity of the land are not put first, the seasonal gleaners operate de facto in a fashion similar to that of the corporations, which use local communities and their landscapes solely as economic colonies. This is something local communities must help seasonal gleaners understand if destructive competition is to be avoided.

Competition

In contrast to the above, there are communities where wealthy people move in, drive up land prices, and effectively take over the town by forcing out the community's original inhabitants. The displaced members are forced to live elsewhere but are allowed to commute from their new homes to their original community, where they may work to serve the wealthy. Whether this happens by default or by design, the effect is the same: trust is irrevocably broken, as is the historical continuity of the community.

Many people would say that all this is simply the way of competition and that makes it okay. But a vision directed solely by competition cannot long endure; it must deplete itself. Continual depletion of natural resources, and with them local communities, is a danger we daily face because we are so overdependent on and mesmerized by competition that it is our predominant model for learning and change.

Although conventional wisdom says there is nothing intrinsically wrong with competition, that it can even be fun and promote invention and daring, we have lost the balance among competition, cooperation, and coordination at precisely the time when we most need to work with one another. Economic competition, which today is being globalized, increasingly pits workers in each enterprise against workers in all enterprises, workers in each ethnic group against workers in all ethnic groups, and workers in each country against workers in all countries.

Economic competition can only destroy social/environmental sustainability—never forge its links. We thus find ourselves oftentimes competing with the very people with whom we need to collaborate, which frequently leads to destructive conflicts over the way in which resources are used and who gets what, how much, and for how long.

RESOLVING CONFLICTS 3

Our culture fosters two myths about conflict. One is that conflict is negative; the other is that conflict must produce a winner and a loser. Neither is true, however, because we are responsible for our own beliefs about conflict. Unfortunately, we too often see conflict as a necessary result of someone else's behavior rather than as a *process* for growth in our own consciousness.

Buckminster Fuller told us that if we really want to understand the principles of the Universe, we must look closely at Nature, where, he said, conflicts result only in change, such as the creation of the Grand Canyon out of a flat plateau. If, therefore, we choose to change our beliefs about conflict, we can change the way we are in conflict and use it to both recognize and create opportunities, because a shift in belief demands a shift in action.

Conflict is thus a choice of behavior based on a particular belief. We resort to destructive conflict in one way or another, at one level or another, because that is what we are taught to do as children. That is how we were taught to cope with change—those circumstances perceived as threatening to our survival.

Looking around the world today, various segments of the global society are blowing themselves to bits and in the process needlessly, recklessly squandering the natural resource base on which they and future generations depend for survival. Children are being ushered into emotionally shattered lives, where their inner poverty will compound the outer poverty they face in the spiritual/cultural/economic chaos of disrupted lives; gutted cities; corrupt, power-hungry governments; and war-torn, fragmented landscapes. These generations may well grow up thinking that hatred and destructive conflict are the norm, which continually fosters the unworkable paradigm of a black-and-white world, in which "I'm right and you're wrong" and "you're either for me or against me."

Is such a future unavoidable? Must we increasingly become a world of victims in which there is no escape from an eternal dysfunctional cycle of abuse and combat? If abuse, and the combat it engenders, are indeed the lot of humanity, then time and history will grind wearily on to the only social outcome possible—the ultimate destruction of human society—taking much of life on Earth with it. Is this the lesson human history is to continue teaching as each day's activities are recorded in the archives of eventide?

We think not. We see the world differently, and this difference is predicated largely on two things: (1) recognizing, accepting, and acting on the notion that destructive conflict is a choice, which means we can choose peaceful ways of resolving differences, and (2) understanding that the peaceful way lies in the art of transformative facilitation,[11] whereby differences are resolved through inner shifts in consciousness.

As fear is transcended, perceived differences—fear's cradle—become novel approaches to the commonalities of a shared vision. In fact, they often become a vision's source of strength.

SOCIAL/ENVIRONMENTAL SUSTAINABILITY AND CONFLICT

Social/environmental sustainability will inevitably demand choices different from those we have heretofore made, which means thinking anew. But "a great many people," as American psychologist William James observed, "think they are thinking when they are merely rearranging their prejudices."

Social/environmental sustainability is the frontier beyond self-centeredness and its stepchild, *destructive* conflict. We specify *destructive* conflict because conflict itself is not necessarily destructive. Conflict can be personally and socially constructive, such as a focused debate on an issue that brings about increased growth in personal and social consciousness. In other instances, one can view conflict as somewhat neutral, such as differences of opinion in which two people amicably agree to disagree. A conflict becomes destructive when it becomes personal, destroys human dignity, degrades an ecosystem's productive capacity, or forecloses options for present and/or future generations.

To change anything, however, we must, through the choices we make, reach beyond where we are, beyond where we feel safe. We must dare to move ahead, even if we do not fully understand where we are going or the price of getting there because we will never have perfect knowledge. And we must become students of processes and let go of our advocacy of positions and embattlements over winning agreement with narrow points of view. This is important because our ever-increasing knowledge rapidly outstrips the ability of our current paradigm, based on old knowledge, to explain the new in terms of the old. Before a community stuck in its old paradigm can move forward, however, it has to resolve its conflicts.

True progress toward an ecologically sound and sustainable environment and a socially just and sustainable culture will be initially expensive in both money and effort, but in the end it not only will be mandated by shifting public values but also will be progressively less expensive over time. The longer we wait, however, the more disastrous becomes the environmental condition and the more expensive and difficult become the necessary social changes.

No biological shortcuts, technological quick fixes, or political hype can mend what is broken. Dramatic, fundamental change is necessary if we are really concerned with sustainably bettering the quality of life—even that of next year. It is not a question of whether we *can* change, but one of whether we *will* change. Change is a choice, a choice of individuals reflected in the collective of society and mirrored in the landscape.

CAN DESTRUCTIVE CONFLICTS BE RESOLVED?

Can destructive environmental conflicts be resolved? Emphatically, yes! But thus far we find only one kind of facilitative approach (or philosophy, if you will) that can accomplish such resolution, the largely ignored *transformative approach,* which brings about a positive shift in one's consciousness and hence in one's moral development.

To resolve destructive environmental conflicts, the process of facilitating a resolution not only must have the greatest and longest-lasting personal and social effect possible but also must be as healing as possible, because outcomes of environmental conflicts are, above all, intergenerational. This means that it is the present generation's responsibility to serve the future, not the future generations' responsibility to serve the present.

However, because people are often unconscious of the motives of their choices, some kind of facilitative process is needed to help resolve destructive conflicts by overcoming blind spots. Destructive conflicts, which are created by the choices people make, can be resolved by electing different choices with resolution so firmly in mind that it naturally leads to a mutually beneficial, shared vision of a future toward which to build.

Sustainable community development as a vision must be planned within the context of a sustainable landscape. This means that a community committed to sustainable development is not seeking some mythical "balance" between its economics and its environment. Rather, it seeks the synergism of ecology and culture, including economy, to promote a healthy, sustainable environment that enriches the lives of all its inhabitants—both human and nonhuman. Sustainable community development gives people a chance to employ such cardinal principles of culture as democracy, beauty, utility, durability, and sustainability while there is still time to approach them.

The resolution of any destructive environmental conflict must culminate in an other-centered shared vision toward which to build if society, as we know it, is to survive the 21st century. This is a critical idea with respect to social/environmental sustainability, because parts are often mistaken for wholes and ideas are often viewed as complete when in fact they are not. Such is often the case when the resolution of a destructive environmental conflict is seen only as the solution of an immediate problem.

RESOLVING DESTRUCTIVE ENVIRONMENTAL CONFLICT

Facilitation is generally understood as an informal process in which a neutral third party, one powerless to impose resolution, helps disputing parties seek a mutually acceptable settlement. As such, facilitation has within itself the unique transformative potential to engender moral growth in people by helping them—in the very midst of conflict—to wrestle with difficult inner and outer circumstances and bridge human differences.

Be that as it may, there is substantial evidence to suggest that today's standard facilitation process focuses largely on problem solving, perhaps even more so than in earlier years. In fact, the unique potential of facilitation to achieve moral transformation is receiving less and less emphasis in practice. This potential is therefore seldom realized, and when it is, it is generally serendipitous, rather than the result of the facilitator's conscious efforts. There is currently a crossroad facing facilitation, one reflecting the movement's two basic approaches: problem solving and transformation.

FACILITATION AT THE CROSSROAD

The problem-solving approach to destructive environmental conflict emphasizes the capacity of facilitation to find solutions that generate mutually acceptable settlements, almost always for the immediate benefit of adult humans, regardless of the effect of the settlement on children or the productive capacity of the environment. Facilitators in this approach often endeavor to influence and direct disputants toward settlement in general and even toward the specific terms of a settlement.

As facilitation has evolved, the problem-solving approach has been increasingly emphasized, to the point where this kind of directed, settlement-oriented facilitation dominates the current movement. The premise of the problem-solving approach is that the most important goal is to maximize the greatest possible satisfaction for individuals engaged in a conflict. But as author Gail Sheehy points out, "Human institutions prepare people for continuity, not for change." To us, therefore, the limitations inherent in the problem-solving approach are precisely its narrowness in scope, rigid focus on quantifiable outcomes, and the increasing attempt to eliminate risk, all symptoms of its growing institutionalization.

Designer Milton Glaser captures well our concern with the uncritical institutionalizing of professionalism in facilitating the resolution of disputes when he says, "Professionalism really means eliminating risk. Once you become good at something, everyone wants you to repeat it over and over again. But the more you eliminate risk, the closer you come to eliminating the act of creative intervention."

In contrast to the problem-solving approach, the transformative approach emphasizes the capacity of facilitation for personal growth, which is embodied in the ability to accept risk. Transformative facilitators therefore concentrate on helping parties empower themselves to define the issues and decide the settlement in their own terms and in their own time through a better understanding of one another's perspective.

Transformative facilitators avoid the directiveness associated with the problem-solving approach. Equally important, transformative facilitators help parties recognize and capitalize on the opportunities for personal growth inherently present in conflict. This does not mean that satisfaction and fairness are unimportant; rather, it means that transformation of human moral awareness and conduct is even more important.

The aim of transformative facilitation is to help parties become better human beings by stimulating moral growth and transforming human character, which results in parties finding genuine solutions to their real problems. In addition, the private, nonjudgmental, noncoercive character of transformative facilitation can provide disputants a safe haven in which to humanize themselves, despite having started out as fierce adversaries. This safety helps people feel and express varying degrees of understanding and concern for one another as they grow toward greater compassion, despite their disagreement.

It is clear that coercion of any kind settles no differences and lays to rest no issues. It only degrades human beings and steals hope from their souls. Therefore, the most important aspect of transformative facilitation is its ability to strengthen people's moral resolve and their ability to handle adverse circumstances beyond the immediate conflict. It therefore transforms society for the better by bringing out the intrinsic good in people, often through their willingness to compromise.

COMPROMISE AND THE POINT OF BALANCE

Facilitation is the way to compromise and the point of balance that resolves conflicts. The *mandorla,* a symbol of unity, is a prototype of conflict resolution that has long been secreted in the gathering dust of medieval Christianity.

A mandorla is two overlapping circles with an almond-shaped area in the middle where the contents of each circle integrate. When thus put together, their areas of overlap and integration—their common root—can be found and perceived opposites can be balanced. And all opposites have a common root because they are, after all, merely different perceptions of the same reality. For example, a glass of water is half full or half empty, depending on one's point of view, but the level of water is the same in either case.

Once the overlap is identified, acknowledged, and accepted, people can begin working collectively, extending the area of overlap and integration. Although the overlap is tiny at first, like the sliver of a new moon, it is a beginning, the first healing of the split between opposites.

With diligent work and the passage of time, the sliver becomes as a quarter moon, then a half moon, and a three-quarter moon, until that point is reached where the two circles become one—a full moon, unity, total healing.

The mandorla as a symbol, a process, and a metaphor seems to fit every social/environmental problem. If, however, we are to map the country of the mandorla to our best advantage, we must treat one another with compassion and justice while we explore the hidden potential of the almond-shaped land of overlap and integration.

A CURRICULUM OF COMPASSION AND JUSTICE

The transformative facilitation process must be as gentle and dignified as possible, which means that it must be a continual lesson in compassion and justice taught through the facilitator's example. All parties must emerge with their dignity intact if anything is to be resolved. It is therefore important to remember that *now* is always the time for compassion and justice, because, as Mahatma Gandhi pointed out, "An eye for an eye only makes the whole world blind." In this sense, facilitation as a democratic process is perhaps at its best when the people involved must continue dealing with one another after the dispute is resolved.

Compassion is the deep feeling of sharing another's suffering, of giving aid and support to another person in her or his time of need (which is the act of forgiving another's perceived trespass), of extending mercy. The essence of compassion is best acknowledged in a French proverb: "To know all is to forgive all." This is but saying that as I do the level best I can in all I do, so does everyone else; so where, therefore, is the judgment? Thus, when we forgive all, when we fix no judgment and place no blame, we have compassion.

Forgiveness, in turn, is to see the fear and the pain out of which another person acts and to extend love as an alternative. "Fear," posited American singer Marian Anderson, "is a disease that eats away at logic and makes...[people] inhuman." To have compassion, therefore, demands far greater courage than does retaliation in any form at any time, because compassion demands that one is responsible for one's own behavior and abnegates the role of victim and/or aggressor.

Although one person cannot experience how another person is feeling, one can ask. If the other person can relax long enough to answer, one may be able to imagine oneself for a moment in the other person's situation and see if one would act any differently under the particular circumstances. If one concludes that he or she would do the same if he or she were in the other person's situation, then it is clearly one's responsibility to forgive rather than the other person's responsibility to change, which means that we must do unto others as we would have them do unto us in any given circumstance.

We have often heard it said that "all's fair in love and war" and that "nothing's fair in life." If this is the way people really feel, perhaps we must shift our attention from fairness to justice, or that which is just, which means being honorable, consistent with the highest morality, and equitable in one's dealings and actions.

One can choose to be just, not only in a moral sense but also in a practical sense. By being just and equitable with others, they are encouraged to be just and equitable in return. It is wise, therefore, to cultivate such kindly behavior for oneself by extending it first to others. And if for some reason the other person is unjust in return, it gives one an opportunity to practice compassion. Being compassionate requires nothing from anyone else—only one's own courage and the wisdom to love enough to forgive. Such compassion is one of the gifts of transformative facilitation, and it is passed forward through communication.

COMMUNICATION 4

We cannot emphasize strongly enough the importance of all aspects of communication in the world today, where disenfranchised human populations are being dramatically shifted around the globe. If the general populace will take the time to become better acquainted with the skills of communication, the world may experience less misunderstanding, conflict, and racism and thus a greater chance to recreate the lost sense of community, mutual caring, and trust in the present and hope for the future.

Emotions and knowledge (a reflection of social experience) are shared through communication, which must be treated with the utmost respect. Just as dishonest or careless communication tells much about the people we are listening to, so too does good communication. Good communication means respect for both listener and speaker, because one must first listen to understand and then speak to be understood.

Communication is perhaps one of the most difficult things we do as human beings, and yet it is simultaneously one of the most important things we do. We are creatures who must share feelings, senses, abstractions, and concrete experiences in order to know and value our existence in relation with one another. Communication is the way we share the essence of our relationships. Our very existence revolves around it, and without it, we have nothing of value.

Language as a way of sharing and coordinating our human values is becoming increasingly important and complex as people of different languages and ethnic backgrounds move into communities or visit them seasonally to share in the harvest of renewable natural resources. It is therefore necessary in considering sustainable community development to address language as a means of communication, including cross-cultural communication.

LANGUAGE AS A TOOL

Although most communication is conveyed in tone of voice, body language, attitude, vibrations, and other energies, we will focus first on the spoken word.[12] Words are symbols for the things we experience; therefore, the more accurately a chosen word builds a bridge to our common ground, the easier it is to get in touch with one another, stay in touch, build trust, and ask for and receive help.

In this sense, semantics is more than quibbling over words; it reveals both our patterns of thought and our consciousness of cause and effect. It is the conveyance of concepts, perceptions, personal truths, trust, and a shared vision for the future. Like every linguistic creation, language can empower or limit, depending on whether we see it as a set of labels describing some preexisting, unchangeable reality or as a medium with which to articulate a new reality—a sustainable way of living together on and with the Earth.

Common bonds can be built, maintained, and strengthened through good communication, which is a clear, concise use of language. Just as any relationship requires sensitive, honest, and open communication to be healthy and grow, so too are relationships in community forged, maintained, and improved when feelings and information are shared accurately, freely, and with tact.

The quality of communication is enhanced if simple rather than complex words are used. Picturesque slang and free-and-easy colloquialisms, if they are appropriate to the subject and if they do not offend the sensibilities of the listeners, can add variety and vividness to the conversation. But substandard English, such as grammatical errors and vulgarisms, not only detracts from one's dignity but also reflects one's attitude toward the listener's intelligence.

If the subject under discussion includes technical terms, one must be sure to define each term clearly and concisely so that all present know exactly what is meant by it. It is also best to use specific rather than general words. In addition, to ensure clarity, one must use sentences of short to medium length because, for most people, the spoken word is often more difficult to grasp (even if English is their native language) than the written word, which can be read over and over and studied.

Quality communication requires constant, consistent practice. It is thus imperative that we continually monitor our words, their meanings, and their usage in our everyday speaking and writing. Every word must be valued, and every word that does not carry its weight must be discarded because good communication comes first and foremost from good thoughts. By our thoughts we privately define and by our actions we publicly declare who and what we are. British psychologist James Allen stated this beautifully:

> A man's [woman's] mind may be likened to a garden, which may be intelligently cultivated or allowed to run wild; but whether cultivated or neglected, it must, and will, bring forth. If no useful seeds are put into it, then an abundance of useless weed seed will fall therein, and will continue to produce their kind.
>
> Just as a gardener cultivates his plot, keeping it free from weeds, and growing the flowers and fruits which he requires, so may a man tend the garden of his mind, weeding out all the wrong, useless, and impure thoughts, and cultivating toward perfection the flowers and fruits of right, useful, and pure thoughts. By pursuing this process, a man sooner or later discovers that he is the master gardener of his soul, the director of his life. He also reveals, within himself, the laws of thought, and understanding, with ever-increasing accuracy, how the thought forces and mind elements operate in the shaping of his character, circumstances, and destiny.[13]

A word spoken is thus the manifestation of a thought, whether positive or negative. Once spoken, it can never be withdrawn, despite an apology, because words are public extensions of our private selves.

Good communication, a prerequisite for both teaching and learning, clears the way for shared, participative ownership of ideas as a means of building relationships. There are, however, a number of obligations that accompany good communication.

Because the right to know is basic to trust in relationships, all parties must have equal access to pertinent information if a misunderstanding is to be avoided. Here, we believe, it is better to err on the side of sharing too much information rather than risk someone being left in the dark.

Everyone has a right to simplicity and clarity in communication and an obligation to communicate simply and clearly. We must go to this length because we owe everyone truth and courtesy, although truth is often uncomfortable and at times a real constraint, and courtesy may be an inconvenience. Nevertheless, it is these qualities that allow communication to educate and liberate us.

One is obliged to practice discrimination in both what one says and what one hears, which means that one must respect one's own language through its careful usage. Muddy language means muddy thinking, and muddy thinking means muddy language. One must therefore always remember that the people who comprise one's audience (which may be freshly transplanted from Latin America, Asia, or somewhere else) may need something special, such as an extraordinary amount of patience and clarity while they struggle to communicate.

Language is the most profound tool we have because it both educates and liberates. Teaching and learning underlie community literacy and action. Community literacy is "why" the process does what it does, and action is "what" it does. With this in mind, we can use language to help the people engaged in relationships to build the freedom of trust. To allow people to build trust, however, communication must be based on sound reasoning, compassion, detachment, and sometimes on silence—but always on the need to be heard.

SILENCE AND THE NEED TO BE HEARD

One must learn to appreciate the power of silence in communication. Most people are profoundly uncomfortable with silence and feel compelled to speak. Silence, when allowed to flow unimpeded through indeterminate seconds and minutes, draws people out, causing them to engage both uncomfortable circumstances and one another. On the other hand, silence (listening) also allows people to be heard.

Although one may not think of it as such, listening is the other half of communication. Communication is a gift of ideas; therefore, a person can give someone else a gift of ideas through speaking only if the other person accepts the gift through listening. The spoken word that falls on consciously "deaf ears" is like a drop of rain evaporating before it reaches the soil. Intolerance of another's ideas belies faith in one's own.

The watchword of listening is empathy, which means imaginative identification with, as opposed to judgment of, a person's thoughts, feelings, life situation, and so on. The more one can empathize with a person, the more she or he feels heard, the greater the bond of trust, and the better one understands the situation. This means, however, actively, consciously listening with a quiet, open mind, without forming a rebuttal while the other person is speaking. Such listening is an act of love, and anything short of it is an act of passive violence.

Not listening is an act of violence because it is a purposeful way of invalidating the feelings—the very existence—of another person.

Everyone needs to be heard and validated as a human being because sharing is the bond of relationship that makes us "real" to ourselves and gives us meaning in the greater context of the Universe. We simply cannot find meaning out of relationship with one another. Therefore, only when we first validate another person through listening as an act of love can that person really hear what we are saying. Only then can we share another's truth. Only then can our gift of ideas touch receptive ears.

All we have in the world as human beings is one another, and all we have to give one another is one another. We are each our own gift to one another and to the world; we have nothing else of value to offer. We cannot give our individual gifts, however, if there is no one to receive them, if there is no one to hear. Therefore, if we listen—really listen—to one another and validate one another's feelings, even if we don't agree, we can begin to resolve our differences before they become disputes. But to listen well and to speak well, it is important to consider the basic elements of communication.

THE BASIC ELEMENTS OF COMMUNICATION

Communication occurs when one person transmits ideas or feelings to another or to a group of people. Its effectiveness depends on the similarity between the information transmitted and that received, a common frame of reference.

The process of communication is composed of at least three elements: (1) the sender—someone speaking, writing, signing, or emitting the silent language of attitude or movement; (2) the symbols used in creating and transmitting the message—sounds of a particular and repetitive form called spoken words, particular and repetitive handcrafted signs called written words, a particular arrangement of musical notes called melody, and facial expression and body language; and (3) the receiver—someone listening to, reading, or observing the symbols. These elements are dynamically interrelated, and that which affects one influences all.

Suppose a boy has something in mind that he wants to convey to a girl. He tries sending his thoughts through the air as intelligent noise for her to pick up with her receivers, her ears. She must then translate the sounds back into her own thoughts that simulate his thoughts as she understands them. And she thinks she knows what he said? He cannot even accurately tell her what he meant because there are seldom words with which to express clearly the nuances of thought. How, for example, can one really say "I love you"? What does that mean? One can feel it, but there simply are no words to describe the feeling.

Communication is thus a complicated, two-way process that is not only dynamic amongst its elements but also reciprocal. If, for instance, a receiver has difficulty understanding the symbols and indicates confusion, the sender may become uncertain and timid, losing confidence in being able to convey ideas. The effectiveness of the communication is thus diminished. On the other hand, when a receiver reacts positively, a sender is encouraged and adds strength and confidence to the message. Let's examine how the three elements work.

Sender

A sender's effectiveness in communication is related to at least three factors. First is facility in using language, which influences the ability to select those symbols that are graphic and meaningful to the receiver.

Second, senders, both consciously and unconsciously, reveal their attitudes toward themselves, toward the ideas they are transmitting, and toward the receivers. These attitudes must be positive if the communication is to be effective. Senders must indicate that they believe their message is important and that there is a need to know the ideas presented.

Third, a successful sender draws on a broad background of personal, accurate, up-to-date, stimulating, and relevant information. A sender must make certain that the ideas and feelings being transmitted are relevant to the receiver. The symbols used must be simple, direct, and to the point. Too often, however, a sender uses imprecise language and/or technical jargon that is nonsense to the receiver and thus impedes effective communication.

Symbols

The most basic level of communication is achieved through simple oral and visual codes. The letters of the alphabet, both spoken and written, constitute such a basic code when translated into words, as do common gestures and facial expressions. But words and gestures only communicate ideas when combined in meaningful wholes: speeches, sign language, sentences, paragraphs, or chapters. Each part is critical to the meaning of the whole.

Ideas must be carefully selected if they are to convey messages that receivers can understand and react to. Ideas must be analyzed to determine which are best suited for starting, carrying, and concluding the communication and which clarify, emphasize, define, limit, or explain the context—all of which form the basis of effective transmission of ideas from the sender to the receiver.

Finally, the development of ideas from simple symbols culminates in the selection of the medium (such as hearing or seeing) best suited for their transmission. In cross-cultural communication, however, a variety of media (hearing, seeing, touching, and at times smelling and tasting) is most effective because it relates to the widest range of experiences and senses.

Receiver

While a basic rule of communication is to be clear, concise, and relevant, communication is a shared responsibility in which the receiver must do his or her best to understand. One knows communication has occurred when receivers react with an understanding that allows them to change their behavior.

To understand the communication process, it helps to appreciate at least three aspects of receivers: their abilities, attitudes, and experiences, which often and in many hidden ways relate to one's ethnic background. First, it is important to discern a receiver's ability to question and comprehend the ideas transmitted. One can encourage a receiver's ability to question and comprehend by providing a safe atmosphere that welcomes such participation.

Second, a receiver's attitude may be one of resistance, willingness, or passivity. Whatever the attitude, one must gain the receiver's attention and retain it. The more varied, interesting, and relevant one is, the more successful one will be in this respect.

Third, a receiver's background, experience, and education (often extremely diverse in a group situation) constitute the frame of reference toward which the communication must be aimed. One must assume the obligation of assessing the receiver's knowledge and of using it as the fundamental guide for effective communication. To get a receiver's reaction, however, one must first reach him or her, and it is in this area that the major barriers to communication are usually found.

BARRIERS TO COMMUNICATION

The nature of language and the way in which it is used often lead to misunderstandings and conflict. These misunderstandings stem primarily from six barriers to effective communication: (1) lack of a common experience or frame of reference, (2) how one approaches life, (3) use of abstractions, (4) inability to transfer experiences from one situation to another, (5) words and language, and (6) intercultural language.

Lack of a Common Experience or Frame of Reference

The lack of a common experience or frame of reference is probably the greatest barrier to effective communication. Although many people believe that words carry meaning in much the same way as a person transports an armful of wood or a pail of water from one place to another, words never carry precisely the same meaning from the mind of the sender to that of the receiver. Words are vehicles of perceptive meaning. They may or may not supply emotional meaning as well. The nature of the response is determined by the receiver's past experiences surrounding the word and the feelings it evokes. And there are words in most languages that simply cannot be translated into another language.

Feelings grant a word its meaning, which is in the receiver's mind and not in the word itself. Since a common frame of reference is basic to communication, words in and of themselves are meaningless. Meaning is engendered when words are somehow linked to one or more shared experiences between the sender and the receiver, albeit the experiences may be interpreted differently. Words are thus merely symbolic representations that correspond to anything to which people apply the symbol—objects, experiences, or feelings.

Thus, a sender must differentiate carefully between the symbols and the things they represent, keeping both in as true a perspective as possible. The truth of a perspective (the interpretation of an experience) is often based to a large degree on one's ethnic background and functionality of one's family of origin. It is also based on the degree to which a person has grown not only beyond his or her own familial background but also in the breadth and depth of his or her individual life experiences. Taken altogether, this translates into generalized personality traits.

How One Approaches Life

In a sense, generalized personality traits are an amalgamation of the mechanisms with which one navigates life. They thus become the essence of one's interpretation of life experiences and the springboard of one's personal capabilities. These traits, which we each possess to a greater or lesser degree, are not cut-and-dried, but rather are overlapping tendencies with varying shades of gray. They are discussed here (with no intent to be pejorative) because they often form substantial barriers to communication.

For example, some people can take ideas seemingly at **random** from any part of a thought system and integrate them; these people have mental processes that instantly change direction, arriving at the desired destination in a nonlinear, intuitive fashion. Others can think only in a **linear sequence**, like the cars of a train; these people have mental processes that crawl along in a plodding fashion, exploring this avenue and that, without assurance of ever reaching a definite conclusion. If the random thinker is also at ease with **abstractions** but the linear-sequence thinker requires **concrete** examples, their attempts to communicate may well be like two ships passing in a dense fog.

Then there is the **introverted** person who appears self-possessed, even aloof. An introverted person processes things internally, navigates life's path more or less alone, and has few friends over a lifetime. An **extroverted** person, on the other hand, is outgoing, mingles easily with other people, requires the presence of people to be happy, processes things through mutual discussion, and has a constant string of friends. An introvert works well alone behind the scenes, whereas an extrovert works well with people out front. In addition, there are four other traits, which can be summed as fatalist, exasperater, appraiser, and relater.

A **fatalist** is the consummate victim who feels powerless in the face of an all-powerful system or life itself and is forever suffering a loss of control. To this person, the operational word is "can't." A fatalist, resigned to his or her lot in life, is often barely functional and requires a tremendous amount of energy, sucked from whomever will give it, to even reach zero on the scale of enthusiasm. Just as soon as the support person stops propping up the fatalist, however, he or she plunges below zero again.

Although the fatalist wants to be rescued, he or she will resist any attempted rescue at any cost. Here one must be wary. The only one who can rescue a person is the person himself or herself. And only he or she knows when he or she is ready for self-rescuing.

Fatalists tend to be good technicians and are most comfortable with simple, clear instructions about which they do not have to think. Having said this, it is critical to understand that fatalists are usually paralyzed by having to bear responsibility. They work well behind the scenes, are usually patient with details, and may even agree to monitor the progress of an activity, provided they do not have to accept any responsibility for its outcome.

An **exasperater**, on the other hand, must be the center of attention and is deeply vested in so being. Here the watchword is control. Some exasperaters go to great lengths to command attention and be in control of whatever they are involved in. They tend, for instance, to be good at "one-liners," know "all" the jokes, be the life of the party, and will argue any and every side of an issue, even changing sides in midstream, rather than acquiesce.

Exasperater personalities are as persistent as a bulldog. Rather than agree, they will say, "Yes, but..." each time someone tries to show them another way of thinking about something or another possible outcome.

It is best to openly and freely acknowledge the exasperater's point of view, the supposed position of power, which does not mean that one necessarily agrees with it. Once exasperaters feel they have exerted their power and have been appropriately recognized, they can usually relax and everyone can get on with the process of communication.

Once an exasperater has an idea in mind, however, he or she becomes impatient for action and, throwing caution to the wind, often barges ahead without getting adequate data and/or listening to other sides of a discussion. On the flip side, if you want to get something done, done well, and completed on time, give it to an exasperater because he or she will move heaven and earth to show off his or her prowess.

While an exasperater often "knows it all," an **appraiser** wants facts, facts, and more facts! An appraiser seems to be uncertain in the world and wants to make sure that all the data are in, examined, weighed, reexamined, and reweighed before any decision is made. Such caution demands much patience because an appraiser often seems to hold the forward motion of meaningful communication in abeyance, regardless of how much data are at hand.

It is prudent, when dealing with an appraiser, to constantly refer to all data that are available and to relate such data to the process and its potential outcome. If data are needed, ask an appraiser to obtain it, and you will likely get the best there is—and lots of it.

Then there is the **relater**, the person who is vitally concerned with what others will think and will go wherever the political wind blows. The relater seldom seems to know who he or she is and seems to have ideas only in relation to their acceptability to others. Such a person changes his or her mind often and gives away his or her power to whoever asks for it.

Since success or failure is not an event but rather the interpretation of an event, successes or failures of relaters are determined by what everyone else thinks, because relaters are constantly comparing themselves to those around them and internalizing what they are told by others. Unfortunately, we usually lose in the end when we compare ourselves to others because we tend to select someone we admire and then find our differences to be deficiencies, even liabilities.

Relaters seem subject to having their feelings or pride hurt easily and often. This is perhaps the major way in which they try to control uncomfortable circumstances because it causes most people around them to "walk on eggshells."

In working with relaters, it is best to refuse to accept their power, even when it is offered. Instead, ask them what they think and how they feel in an effort to draw them out. Done gently and patiently, this can work quite well.

Relaters are generally excellent with public relations because they are sensitive to how others feel and work very hard to win approval. They thus have a good sense of how to market an idea.

There are also **piece thinkers**, the people who tend to focus on individual pieces of a system, or its perceived products, in isolation of the system itself. **Systems thinkers**, on the other hand, tend more toward a systems approach to thinking. A person oriented to seeing only the economically desirable pieces of a system seldom accepts that removing a perceived desirable or undesirable piece can or will negatively affect the productive capacity of the system as a whole. This person's response typically is "Show me; I'll believe it when I see it."

In contrast, a systems thinker sees the whole in each piece and is therefore concerned about tinkering willy-nilly with the pieces because he or she knows such tinkering might inadvertently upset the desirable function of the system as a whole. A systems thinker is also likely to see himself or herself as an inseparable part of the system, whereas a piece thinker normally sets himself or herself apart from and above the system. A systems thinker is willing to focus on transcending the issue in whatever way is necessary to frame a vision for the good of the future.

The more a person is a piece thinker, the more reticent he or she is to change. This type of individual sees change as a condition to be avoided because he or she feels a greater sense of security in the known elements of the status quo, especially when money is involved. But, as Helen Keller once said: "Security is mostly a superstition. It does not exist in Nature. Life is either a daring adventure or nothing." Conversely, the more of a systems thinker a person is, the more likely he or she is to agree with Helen Keller and risk change on the strength of its unseen possibilities.

A piece thinker is likely to be very much concerned with land ownership and the rights of private property and wants as much free rein as possible to do as he or she pleases on his or her property, at times without regard for the consequences for future generations. The more of a piece thinker a person is, the greater the tendency to place primacy on people of one's own race, creed, or religion, as well as on one's own personal needs, however they are perceived. The more of a piece thinker a person is, the greater the tendency to disregard other races, creeds, or religions, as well as nonhumans and the sustainable capacity of the land. Also, the more of a piece thinker a person is, the more black and white one's thinking tends to be, as illustrated in the following example:

> The wimpy [sic] comments by Mike Mitchel in Sunday's "Rural Issues" were disturbing. As the head "honcho" and decision-maker for a BLM [Bureau of Land Management] office, he said things like, "We just follow the regulations and enforce them...." Also, "We have our regulations and have no choice."
>
> That's typical bureaucratic arrogance, and a cop-out. Those regulations didn't come down the mountain on stone tablets. They are the product of a well-funded lobby in Washington, D.C., that represents those who are "saving us" from the horrible ranchers, miners and farmers of Nevada.
>
> He says the land will restore itself in 15 or 20 years if we change grazing practices. Restore itself for what? So some manicured marshmallow-butt from Washington can start up a cattle ranch on abandoned land? Get real! Nevada ranchers are on the land now! The BLM should help them do what they do best, or get the hell out of the way. I'm not to [sic] smart, but I recognize typical Sierra Club rhetoric when I hear it.
>
> As the song goes: When will they ever learn?[14]

A systems thinker, on the other hand, is likely to be concerned about the welfare of others, including those of the future and their nonhuman counterparts. Systems thinkers also tend to be concerned with the health and welfare of Planet Earth in the present as well as the future. And they more readily accept shades of gray in their thinking than do piece thinkers.

There are still other generalizations that can be made, such as people who are visually oriented as opposed to those who respond to sound or touch. In addition, these traits come in a variety of combinations, which indicates how different and complex people can be in response to their life experiences. These differences and complexities naturally carry over into people's patterns of communication. None of these patterns is better than any other as far as communication is concerned; each is only different and needs to be understood, which brings us to the use of abstractions.

As if this is not enough, there are other behavioral traits that affect communication, such as **passive–aggressive behavior**. Passive–aggressive behavior, as a personality trait, is usually disruptive in that such a person is prone to expressing her or his aggression through such passive means as purposefully being late to a meeting that she or he does not want to attend but feels obligated to do so. Other passive–aggressive behavior includes withdrawing from a discussion as a way to control a circumstance by purposefully disrupting communication or refusing to answer a direct question, again being purposefully disruptive. Although most people probably think of passive–aggressive behavior solely as a trait of individuals, it can also be a cultural trait.

Use of Abstractions

Concrete words refer to objects a person can directly experience. Abstract words, on the other hand, represent ideas that cannot be experienced directly. They are shorthand symbols used to sum up vast areas of experience or concepts that reach into the trackless time of the future. Albeit they are convenient and useful, abstractions can lead to misunderstandings.

The danger of using abstractions is that they may evoke an amorphous generality in the receiver's mind and not the specific item of experience the sender intended. The receiver has no way of knowing what experiences the sender intends an abstraction to include, especially if one is dealing with cross-cultural communication. For example, it is common practice to use such abstract terms as "proper method" or "shorter than," but these terms alone fail to convey the sender's intent. What exactly is the "proper method"? "Shorter than" what?

When abstractions are used, they must be linked to specific experiences through examples, analogies, and illustrations. It is even better to use, as much as possible, simple, concrete words with specific meanings. In this way, the sender gains greater control of the images produced in the receiver's mind, and language becomes a more effective tool.

When dealing with communities in relationship to their landscapes, especially cross-cultural communities or cross-cultural community interactions, it is advisable to get people into the field, where they can physically wander through the area and discuss it. Abstractions can thus be transformed into concrete examples, which one can experience through sight, touch, smell, sound, taste, or any useful combination thereof.

Inability to Transfer Experiences from One Situation to Another

Another major barrier to communication is the inability to transfer the outcomes of experience from one kind of situation to another; again, this is especially true when people of different nationalities and ethnic backgrounds come together in discussing contentious issues. The potential ability to transfer results of experiences from here to there is influenced by the breadth of one's personal experiences. Every nationality or ethnic group represents a vast array of experiences, some broad, others narrow, all different.

Experiential transfer, however, is critical to understanding how ecosystems and their interconnected, interactive components function, including the bridge between a community and its surrounding environment. It is also a necessary ability in dealing with contentious issues between a community's sustainability and that of its landscape to be able to show how potential outcomes can be projected to a variety of possible future conditions.

When a person cannot make such transfers for lack of the necessary frame of reference, he or she will find the ideas to be abstractions, whereas others with the required experience will feel them to be concrete examples, based on their accumulated knowledge. This is where analogies are useful.

To make sure that an analogy will be understood, one must ask the person or people to whom one is speaking if he/she or they are familiar with the concrete example that one proposes to use in helping to extend the frame of reference to include the abstraction. If, for instance, one is talking about the value of understanding how the various components of an ecosystem interact as a basis for the system's apparent stability, one can use simple examples, as follows:

1. What happens when just one part is removed? Let's say a helicopter crashes and people are injured. The immediate question is what happened and why.

A helicopter has a great variety of pieces with a wide range of sizes. Suppose the particular problem here was with the engine, which is held together by many nuts and bolts. Each nut and bolt has a small sideways hole through it so that a tiny "safety wire" can be inserted; the ends are twisted together to prevent the tremendous vibration created by a running engine from loosening and working the nut off the bolt.

The helicopter crashed because a mechanic forgot to replace one tiny safety wire that kept the lateral control assembly together. A nut vibrated off its bolt, the helicopter lost its stability, and the pilot lost control. All this was caused by one missing piece that altered the entire functional dynamics of the aircraft. The engine had been "simplified" by one piece—a small length of wire.

Which piece was the most important part of the helicopter? The point is that each part (structural diversity) has a corresponding relationship (functional diversity) with every other part. They provide stability only by working together within the limits of their designed purpose.

2. What happens when a process is "simplified"? Suppose the newly elected mayor of a city whose budget is overspent guarantees to balance the budget; all that is necessary, in a simplistic sense, is to eliminate some services whose total budgets add up to the overexpenditure. A "simplistic" approach is used here because it is not quite that simple. What would happen, for example, if all police and fire services were eliminated? Would it make a difference, if the price were the same and the budget could still be balanced, if garbage collection were eliminated instead?

The trouble with such a simplistic view is in looking only at the cost of and not the function performed by the service. The diversity of the city is being simplified by removing one or two pieces or services, without paying attention to the functions performed by those services. To remove a piece of the whole may be acceptable, provided we know which piece is being removed, what it does, and what effect the loss of its function will have on the stability of the system as a whole.

Once it is certain that an analogy is understood, one can help another person transfer the concept to the abstraction. As the principle of transfer becomes clear, the abstraction begins to take on the qualities of a concrete idea, which usually dissolves this barrier to communication. There is, however, yet another potential barrier—the use of words and the language they compose.

Words and Language

Foreign languages can also act as barriers to understanding, even with an interpreter. Although it seems obvious that a foreign language that one cannot speak is a barrier, there is more to it than that. Even when one learns to speak a different language, that is no guarantee that one will be readily understood or can readily understand because, although spoken and written language is composed of words, language as a whole is more than mere words, some of which cannot even be translated.

Words

The more you work with sustainable community development, the more you will become aware of how important words and their meanings—semantics—are.[15] These symbols are the primary means through which we converse with one another, which makes it important to spend some time thinking about the relationship between the words we use and the ideas we want to express.

Have you ever wondered why you can think so clearly and easily about an idea, yet have such a difficult time finding the right words to convey what you want to say to others? Although we both think and speak in words, each person's vocabulary is unique.

Consider that one person translates silent thoughts into spoken words, which are sent as sounds through space on air waves. Another person must receive the sounds through his or her receivers (ears). That person must then translate them back into words, based on his or her understanding of what the sounds represent. He or she must then translate the words into thoughts, based on his or her understanding of what the words mean.

It is most difficult, therefore, to translate our thoughts and feelings into a spoken language that others can understand. This is especially true when language cannot accurately convey feelings because each person has a different sense of what a particular word means, a sense that may be derived from different backgrounds and interests. Nevertheless, we must be careful that the words we choose when speaking are understood, as much as possible, in the same way by those people hearing us.

Words, it is well to remember, have a variety of functions, which we all know but rarely think about. We are thus surprised when something in a conversation goes awry. Let's consider five functions of words.

1. Words are used as labels to classify things, such as your name, which means you, or objects, such as a table. In the latter case, the purpose of the word is to erect a mental image of the object the speaker has in mind.
2. Words as labels are sometimes misleading because while they are close approximations of true meanings, they are sometimes mistaken for the real thing. In one way or another, therefore, a word can never tell the whole story.
3. Words are often misinterpreted by speaker and/or listener. A speaker using the wrong word because it sounds close to the correct one may elicit a chuckle, but a listener misinterprets words in silence and may magnify the error through explaining his or her understanding to others. This kind of misinterpretation may even be compounded when translating the meanings of words from one language into another language through a translator. If the translator misinterprets a word, the misinterpretation is passed on to the listener, who hears an incorrect translation and, being ignorant of the error, passes it along.
4. Words are sometimes "loaded," in that they bring up strong emotions in one person when to another person they seem completely innocuous. Because words sometimes have emotional tags attached to them, particularly between cultures, people's sensitivity to one another is critically important, especially if one is selected as a community's spokesperson.
5. Words, by the feelings they convey, can make attitudes permanent. Everyone has momentary feelings of admiration or dislike for something or someone. By putting these feelings into words, we tend to fix them permanently in our minds.

In addition to the above-listed functions of words, some words are misleading and at times harmful to clear communication. These words include "all," "is," jargon, and discussion-killing words, and we must be careful how we use them.

1. "All" may be a false generalization. Before one uses "all," one must be sure it fits precisely the statement being made. If it does not precisely fit the statement being made, the use of "all" may arouse anger and/or threaten the speaker's credibility while blunting the ability of both the speaker and listener(s) to think discriminatively.
2. "Is" may lead to a false identity. Linking items together by identity, using "is" statements, must be done carefully to avoid misleading or possibly even insulting the listener. If, for example, you say someone is this or is that, you are making a judgment about that person without really knowing, because all you can see is *what* the person does, not *why* he or she does it. Use of "is," however, assumes that you know both the "what" and the "why," when in fact you don't.
3. Jargon, inappropriately used, can totally befuddle a single listener or a large audience. It is important to use words appropriately if your intent is to be understood, so you in turn can understand.
4. Discussion-killing words come from a person stating his or her opinion flatly, which leaves no room for discussion. Examples of such flat statements are: "The experts agree..." or "The fact is..." or "Everyone knows...." These kinds of statements will stifle a discussion and often put a person with a different point of view on the defensive with no way to be heard.

Although good speech and good writing have much in common, each has a different emphasis. Spoken language must be immediately understandable to the person listening because the listener does not have the opportunity for contemplation and reflection that written language both allows and often encourages.

Spoken words are a one-shot affair; once spoken, they are gone from you, and you can never retract them, although you may wish you could and try to do so. It is imperative, therefore, that you think *before* you speak, because an unkind word, an unclear word, an inappropriate, incorrect, or unfamiliar word can lose a listener. And once your listener is lost, your point is most likely lost also.

In order to give a single listener or a whole audience the best chance to understand what you are saying, it is necessary to construct what you are going to say in an oral style, for which there are three helpful qualities: (1) clarity, (2) appropriateness, and (3) vividness.

Clarity is the first quality of good spoken language. An oral presentation is clearest when short, simple sentences are used, with frequent pauses in the sequence of ideas for explanations and summaries. Spoken language is more informal than written language and tends to be repetitious, which means that by using parallel constructions or repeated phrases, one's ideas become clear, whereas repetition is usually eliminated from written language.

The best words are not only simple but also specific and concrete. The more specific one's words are, the less room there is for those listening to become confused. If, for example, you want your listeners to picture a tomato, use the specific term "tomato," not the general term "vegetable."

Appropriateness is the second quality. Unlike a writer, who can only guess who the readers will be, a speaker has the advantage of seeing a real audience. This means the speaker can deliberately choose his or her words and relate them directly to the listeners.

Beyond the obvious courtesy of using civil language and avoiding vulgarity, inappropriate jargon, and excessive slang, it is a good idea to use personal pronouns whenever possible. The use of such personal pronouns as "you," "we," and "our" establishes the speaker's identity with and shows his or her interest in the audience. The importance of appropriate language is to use words and ideas that place both speaker and audience at a common vantage point, which allows the audience to identify with the speaker.

Vividness is the third quality of spoken language. To make the spoken language alive and memorable, use words that are alive—words that convey images and/or feelings that will stick in the minds of the listeners. The more examples, stories, word pictures, and words of action and feeling one uses, the more vivid and memorable one's spoken language will be and the more effectively one will be able to communicate.

It also helps to group adjectives and points in threes or fours. Expressed in threes or fours, thoughts not only have a pleasant rhythm but also are dramatic and memorable. For example, the following sentence attracts attention and is easy to remember: "All I have to give you of value are my love, my trust, my respect, and the benefit of my experience." And finally, this sentence is an example of individual words becoming language.

Language

Language is an expression through words of a whole system of logic, a whole way of thinking. If one is conversing with another person who culturally has a similar frame of reference, such as an American speaking with a German or a Slovakian, the logic expressed through the conversation is seldom a barrier to communication.

But if an American is conversing with a person from Nepal, India, Sri Lanka, Japan, or China, the logic expressed by the American is very different from that of the others. On the other hand, there are more similarities in thinking among people from these cultures than there are between them and an American. Such differences must be taken into account in cross-cultural communication, especially when metaphors, idioms, and analogies are used because they are likely to miss the mark if one is not familiar with the philosophical underpinnings of the other culture(s).

Although a skilled interpreter/translator can be of enormous help as a cultural liaison, there are things one can do to make the task of the interpreter/translator much easier: (1) if you are delivering a speech, give the interpreter/translator a copy of what you are going to say well ahead of time, which allows prior translation and time for any questions of clarification; (2) if technical words are unavoidable, or if proper or scientific names are used, go over them with the interpreter/translator prior to the meeting; (3) enunciate precisely; (4) speak slowly and clearly with frequent pauses to allow the interpreter/translator to keep apace, (5) check the other person's understanding by requesting that he or she paraphrase important points; and (6) if the other person does actually seem to understand, ask him or her to give you an example from his or her own experience as a matter of your own edification and as a final check of your mutual understanding.

Communication (particularly language) is our most important tool as human beings, and there are things we can do to improve communication among cultures, as some of the following ideas indicate.

Intercultural Language

By intercultural language, we mean that members of a local community who of necessity deal with people of other cultures must appreciate how those people may perceive them.[16] Here it is important to understand how the meaning of the term "culture" has changed over time. The original meaning of "culture," which comes from the Latin *cultus*, was reverential homage. It then came to describe the practice of cultivating the soil and what grows therein and was later extended to cultivating and refining one's mind and manners. Finally, culture, by the 19th century, stood for the intellectual and aesthetic side of civilization.[17]

With this in mind, be aware that the moment you, or anyone else, join people of another culture, you are being stereotyped, and you are in turn stereotyping. Whether it is right or wrong is beside the point. It happens and you must be cognizant of it because how people from other cultures perceive you is how they will likely think about everyone within the community. It is good, therefore, to surprise them and break the stereotype by learning something about their culture and country.

Most non-Americans see Americans as geographically and culturally illiterate, and with good reason. But this is easily remedied by finding out which cultures are represented by the people who have moved into the landscape surrounding your community to harvest seasonal resources. Then get out an atlas and look up their country; study its geographical and social features. Read about it, and then, with some knowledge, you can ask them intelligent questions about their homeland and customs.

Additionally, most non-English-speaking people think Americans can only speak English. Clearly, non-English-speaking people struggle with English, which is a most difficult language to learn. Knowing this, a bond of trust can be initiated by learning to speak a few words of someone else's language, such as "hello," "good-bye," "how are you," and above all, "thank you." This is important because effort and friendliness are universally understood.

Having said this, let's see how some other people see us as Anglo-Americans. We are seen as open and direct in our communications, which in some cultures is deemed blunt, clumsy, and pushy and can easily offend people from Latin America and Asia. We, in our bluntness, miss the nuances of what is really being conveyed.

Rarely, for example, will a Mexican or Japanese answer "yes" or "no." When a Japanese says "yes," it only means that he or she has heard you. When a Japanese says "that will be difficult," he or she means "no." You must therefore judge the response to a question put to a person from Latin America or Asia by the context of what is being said, not by what is literally said. Put differently, you may understand the words used but not what the person *means* by those words.

Americans are often seen as ill-prepared "lone cowboys" who always want to do everything in a hurry because "time is money." Such impatience can cause community leaders to make premature concessions to more patient peoples when negotiating such things as harvest quotas for locally important seasonal resources. It is therefore wise when dealing with Asians to do as Asians tend to do—engage a team to share the work and ideas, even if one person speaks for the community.

Although individualism is good, even necessary, in the embryonic stages of an endeavor, it must blend into teamwork in times of environmental or social crisis. Setting aside egos and accepting points of view as negotiable differences while striving for the common good over the long term is necessary for teamwork. Unyielding individualism represents a narrowness of thinking that prevents cooperation, possibility thinking, and the resolution of issues. Teamwork demands the utmost personal discipline of a true democracy, which is the common denominator for lasting success in any social endeavor.

Because of our "lone cowboy" attitude, we are often seen not only as culturally myopic and arrogant about our nationality but also as being interested primarily in the short-term deal (again, time is money, so let's get on with it) as opposed to the long-term relationship. Asians, on the other hand, tend to be profoundly oriented toward long-term relationships built on personal trust, in which the agreement is the starting point, not the final solution. Under these circumstances, a community is wise to build a long-term, mutually beneficial relationship based on trust and understanding, even if the short term is not as cut-and-dried as an Anglo-American might wish.

We also emphasize the content of our discussions instead of relationships. After an exchange of pleasantries, for example, we want to get right down to business in a logical, factual, legalistic fashion, and it is our insistence on long legal contracts, as a result of our increasingly litigious culture, that gives rise to the perception that we trust neither ourselves nor anyone else.

This behavior, which may be seen as an affront to both friendship and trust, especially among peoples from the Middle East and Latin America, can easily turn off people from a culture in which the main emphasis is placed on interpersonal relationships as a basis of mutual trust. If, therefore, a contract is necessary between a community and people of another culture, make it simple, short, and conversational. No court has thrown out a contract because it was too easy to understand.

Part of the problem is that we are not flexible enough to approach people of other cultures from their vantage point. For example, we often refuse to learn—or even attempt to learn—the language of people from another culture with whom we are dealing, but that is only a choice.

In general, it would be wise for at least the community leaders to make an honest effort to learn about the culture and language of the people with whom they are in contact. We say this because social/environmental sustainability, and hence sustainable community development, is a long-term affair built on making friends and nurturing trusting relationships that will last for years. It is good counsel, therefore, to slow down and listen more than you speak because the goodwill will serve you well over time. It is prudent to begin now.

In addition to spoken and written language and the behaviors associated with it, there is the matter of body language, which is a system of communication based on hand gestures, facial expressions, body postures, and movements that people of all countries and cultures employ daily in their interpersonal relationships. Although we Anglo-Americans tend to think our body language is universal, that is not necessarily true, albeit a clenched fist or a soft warm smile is widely understood. But it is a dangerous mistake to think that all our gestures are global in their meaning because the meaning often changes from region to region and country to country.[18]

To illustrate, you are probably familiar with the American sign for "okay," where a ring shape is made with the thumb and forefinger. Instead of meaning "okay" or "excellent" in France, however, it means something is a big "zero," worthless. In Japan, on the other hand, it indicates money, and in other countries it may indicate a foul obscenity.

Cultures also typically have distances at which they stand when conversing face to face. Consider, for example, a British citizen, whose normal standing distance is at arm's length, talking to an Arab, whose standing distance is half an arm's length. The Arab advances closely, which causes the Briton to feel threatened and retreat until he or she is once again at ease. Now the Arab feels rejected and thus advances again, causing the Briton to retreat again, and so it goes, each knowing something is wrong but neither understanding what it is. The same thing happens when British people talk to Asian Indians and Nepalese. Americans, on the other hand, seem, for the most part, to have little or no sense of interpersonal boundaries and are simply in everybody's face.

Eye contact is another frequently misunderstood language. We in the United States are taught that it is not only polite but also a sign of being in control of oneself to look the other person "straight in the eye" when speaking to him or her. But in other countries, such as those in Asia or the Middle East, this direct, unwavering gaze can be taken as an affront or even a threat.

And then there is the bow. With the Japanese, for example, it is critical to understand the language of the bow and to use it correctly. Bowing too much (too low), which is a common Anglo-American failing, is just as embarrassing to the Japanese as bowing too little.

Unless we take the time to learn the nuances of the body language of the people with whom we are dealing (even if we have to ask to understand), our efforts to be polite may come across as brusque and embarrassing or even downright offensive. Remember, body language can often speak louder than words, even if you do not intend it to do so. It is wise, therefore, to find a person in the other culture who both understands some English and is willing to act as a consultant in teaching you the proper etiquette of his or her culture. Although proper cultural etiquette can go a long way toward productive communication, there also are things we can do to improve our use of language within our own American culture.

GOOD COMMUNICATION

People who speak well truly care about their audience, be it a chamber of commerce or the general populace in a public meeting. They show it by being well prepared, knowledgeable, helpful, and enthusiastic about their message. They also understand that it is easier to treat one another as distant abstractions when we do not get to know one another.

Although speaking well takes practice, a few simple things can improve the skill of one's communication: (1) respect the audience, (2) give the audience something to grasp, (3) pace yourself, (4) make eye contact, and (5) accept controversy as part of personal growth and thus of democracy.

Respect the Audience

A considerate speaker (someone speaking in public) understands that the audience (even if it is just one person) has just as much at stake as does the speaker. A speaker who truly respects the audience is so focused on meeting the needs of those listening that he or she is too busy to focus on himself or herself.

Prepare yourself to speak by analyzing the audience as best you can to determine how knowledgeable they might be about the subject of your presentation and what their attitude might be toward it. This will help you figure out how to present your message for the best effect.

If you use reference materials, make sure they are well designed, clear, to the point, and large enough to be seen from the back of the room. Keep the visuals simple; avoid clutter. Few things are more distracting to an audience than poor visuals, especially for those people who are visually oriented in learning, which will be the majority of your audience. With this in mind, paper copies of your visuals would make excellent handouts.

If you are to speak at a meeting, especially in a formal sense, review the room and find out how the audiovisual equipment works prior to the beginning of the meeting. Ask who will introduce you and find that person. Give the chairperson a written introduction that establishes your credentials and the purpose of your presentation. Also find out where you are to sit before and after your presentation.

Being at your best physically and mentally is a clear sign that you respect the audience. Although being emotionally at your best is sometimes beyond your control, you can give a good performance regardless of what else is going on in your life if you are well prepared.

Remember, people don't care how much you know until first they know how much you care about them. One good way to accomplish this is to accept the people in your audience where they are mentally with respect to the information they have. If you assume they know more than they really do, you may come across as though you are trying to force them to be where you want them to be, in which case they will most likely feel you are talking down to them and will thus feel humiliated, which makes it virtually impossible for them to listen to, much less hear, what you say.

Thus, while you may not like every subject that you may have to address in planning for sustainable community development, you must be authentic in your caring about the audience. And even though what you have to say may be hard for some people to hear, end every presentation on the most optimistic note you possibly can. People are badly in need of hope!

One way to lighten a potentially heavy subject is with humor, provided your personality is suited to this. Those for whom this is not the case are wise to leave most kinds of humor to others. You can, however, always poke fun at yourself or share some of your own foibles, which makes you appear more human and is a great way to build and maintain rapport with your audience.

Although many people use jokes as a form of humor, it is not advised in building toward sustainable community because *a joke is almost always at someone or something else's expense,* which comes across as divisive. This divisiveness is the antithesis of the inclusivity that is part and parcel of sustainable community development.

Give the Audience Something to Grasp

Information, to be effective, must lead to understanding. To ensure that your message leads in that direction, it is important to give your audience well-constructed mental handles to grasp along the way.

Consider, for example, that only one part of learning is made up of information; the other is the context within which to fit the information. This might be likened to a road map, which contains much information on a flat, white background. In and of itself, it tells one little about the terrain through which the roads pass. If, however, this same information is put on a contour map, then the roads are depicted within the context of the surrounding landscape and the twists and turns in the roads make sense.

To be both heard and understood, we must, in our communication, present the audience not only with a verbal map of where we are going (when, how, and why) but also with a mental contextual map, such as time lines (before-and-after or then-and-now comparisons), so they can track the journey and better understand the nuances of the message. This can best be done with an organizational scheme.

Once you have figured out what kind of mind map you want to present, the next step is to determine how to organize your presentation. There are five basic ways to do so: by category, time, location, alphabet, and continuum.

For example, when discussing sustainable community development, you could organize by subject (category), sequence (alphabetically), season of year (time), place of implementation (location), or protocol for monitoring (continuum). To accomplish this, ask yourself: To which category does my material logically lend itself? What organization will make the most sense to my listeners and help them retrieve the information at will?

There is, however, one caution: beware of information overload. Most people can handle three major themes. Beyond that, you will lose much of the audience. The symptoms of imparting too much information to your audience include irritability; hostility toward you, the speaker; confusion; anxiety; and, finally, just tuning out.

Your task, therefore, is to deliver the information at the right pace, in the right amount, and well correlated with common everyday experiences to foster familiarity with the concepts. The more connections you can make between the known and unknown, the less perplexing and intimidating (the more acceptable) new information becomes. This is the best way to make sure your listeners grasp, as best they are able, the main points for recall. And remember, a picture (or concrete experience) is still worth a thousand words.

Some people in your audience may learn best by hearing what you say and others by picturing your words or through the visuals you use. Still others need hands-on experience to understand. This brings to mind a Chinese proverb: I hear and I forget; I see and I remember; I do and I understand. Therefore, to the extent that hearing, seeing, and hands-on experience can be employed when communicating about sustainable community development, people not only will better understand, learn, and remember what you say but also are more likely to become actively involved in the process.

The more knowledgeable you are about your subject or the more frequently you present it, the more susceptible you are to what may be called "the disease of familiarity." You must never assume, however, that just because something seems obvious to you, your audience is either mentally or emotionally where you are. The majority may not have a clue in terms of what you are talking about.

It is wise, therefore, to put yourself in the place of your listeners and thereby attempt to anticipate their questions and concerns. Try to identify which aspects of your material are the most complex or difficult to grasp, and as you proceed, watch for blank stares or expressions of confusion. If you see such expressions, stop and clarify the point before resuming the discussion.

End your session by recapping the major points and their interrelationships; if time permits, give the listeners an opportunity to summarize what they have learned in their own words. It is a good practice to close with a personal reflection, a thought-provoking quotation, or a challenging question that is likely to linger in your listeners' minds well after they depart.

Pace Yourself

Be sensitive to the gap in knowledge between you and your audience and to their preferences; some prefer a slower pace than others. The response of your audience will give you a good idea of how well you are pacing yourself in presenting the information. If people look puzzled, slow down and clarify; if they look sleepy, speed up.

But remember that pace is not speed. Pace is the forward movement of your message at the *appropriate rate,* which means that if you have a time limit, you must plan for it carefully and respect it. In the end, it is a well-paced (and when possible interactive) session that allows participants to get the most out of your presentation.

Make Eye Contact

Connection with your audience is made through eye contact. No eye contact, no connection.

It is essential, therefore, that you look at people, not past them, when you speak. Eye contact is important whether you are speaking to a large group or just one person. It is possibly even more important when you are engaged with a single person because it is said that the eyes are the windows to a person's soul. As a rule of thumb, when you speak to people, focus on their eyes long enough to note the color.

Accept Controversy as Part of Personal Growth and Thus of Democracy

To say anything worthwhile is to risk offending someone. In this sense, most memorable speeches have been controversial.

Be that as it may, John Stuart Mill was of the opinion that unconventional views needed airing, even if they were known to be wrong. Listening to opinion contrary to our own, thought Mill, draws us closer to the truth because the prevailing wisdom on any subject is rarely the whole truth. Only through a "collision of adverse opinion," he said, could the rest of the truth be extracted.

To extract such truth, the world is in great need of people who can articulate deeply held beliefs not only with conviction and passion but also with reason, civility, and grace. The ability and skill to speak both persuasively and high-mindedly on a controversial topic are thus of supreme value because a healthy democracy, which embodies the very foundation of sustainable community development, depends on leaders who can lead without inflaming the opposition with pettiness.

In addition to the above speaking skills, there are two other points concerning communication that are important in building toward sustainable community development: (1) remember the name of the person to whom you are introduced and (2) treat everyone as equals.

Remember the Name of the Person to Whom You Are Introduced

Paying full attention during new introductions is the key to retaining names in your memory. Take the time to focus fully on anyone you are about to meet. The time to recognize a potentially important contact is before you forget the name, not afterward.

Therefore, give your full attention to the introduction. Make the person feel important by making the introduction important to you. Listen carefully and repeat the name immediately, which not only helps you to reinforce the name in your memory but also gives you the opportunity to rehearse and/or double-check difficult pronunciations. Use the name frequently during your conversation.

If perchance you forget a name or did not hear it clearly the first time, ask. Most people prefer to repeat their name rather than correct a wrong guess. But do not make excuses, because an excuse can make even the most important person feel insignificant in your eyes.

Treat Everyone as Equals

One thing most of us probably do not think about is how to treat people with disabilities.[19] We are thus likely to be uninformed or have information that is both inadequate and incorrect about disabled people and how to communicate with them in a way that protects their dignity. The following are a few simple courtesies that can be extended:

1. Treat adults as adults, which means abandoning the concept of "special," because it usually means segregated. Instead, think equal, and call a person by his or her first name only if you are extending this familiarity to everyone present.
2. Avoid regarding and treating someone as heroic simply because he or she has a disability. People with disabilities are not endowed with greater courage than people without disabilities, but rather deserve equal treatment and opportunity.

3. By the same token, do not patronize a person in a wheelchair by patting him or her on the head. Reserve this sign of affection for children, even if the head of the person in the wheelchair is temptingly at about the same height as a child's.
4. When speaking with a person who has a hearing impairment, it is important to keep your hands away from your face. You must also make sure the lighting is adequate for the person to clearly see your face, and you must eliminate as much background noise as possible.
5. Introduce yourself and others in a normal tone of voice to a person with a visual impairment, and be sure to use the person's name so the individual will know you are addressing him or her. Advise the person before you move away.
6. Offer to assist a person with a disability, but wait until your offer is accepted and then listen to any instruction the person may want to give *before* you actually help.
7. Protect the personal space and privacy of a disabled person. It is both insensitive and rude to ask about matters of intimacy and income.
8. Speak directly to a disabled person rather than through a companion who may be present.
9. Sit, squat, or kneel if your conversation with a person in a wheelchair lasts more than a few minutes; this will spare both of you a stiff neck.
10. Consider that a person with a disability might need extra time, and let him or her set the pace in moving and talking.
11. Do not call, talk to, feed, pet, or even make eye contact with a guide or service dog in harness. The animal is working, and a distraction could place both the dog and owner in danger.

The sensitivity with which we treat one another hinges on good communication because it is through language that we are able to share with one another. Removing barriers to linguistic communication is therefore critically important if we are to understand and trust one another well enough to build toward social/environmental sustainability, which if nothing else is a cooperative venture. But first, we must learn something about the language of economics on which community sustainability and community stability are partly based.

5

BEGINNING TO THINK ABOUT ECONOMICS AND SUSTAINABILITY

Before one can begin to think about relationship, one must have some understanding of the language of relationship and its philosophical underpinnings. This is especially true when dealing with two seemingly opposing disciplines that appear to be mutually exclusive along traditional battle lines: economics and environmental sustainability.

In dealing with economics in this book, we have done our best to humanize what has developed into a very depersonalized discipline. Despite our best attempts, however, the chapters on economics may appear to be written in a different style than the rest of the book.

UNDERSTANDING SOME ECONOMIC CONCEPTS

For our discussion of economics to end up with holistic and connected conclusions, we need to start with some linear and hierarchical definitions. A true revolutionary would note the irony of using tools from the old order to establish a new one. The old tools with which we begin can be thought of as notations in economic "shorthand."

WNRCP Hierarchy

First, for convenience in beginning to address the relationships between economics and sustainability, it is necessary to identify (in descending order) five levels at which it is analytically possible to think about sustainability: world, national, regional, community, and personal. At the risk of using an acronym that looks like the call letters of some local radio station, we call this the **WNRCP hierarchy**.

EVE Dilemma

Second, in most economic treatises that in any way refer to both economic and environmental issues, some sort of assumption is generally made about a supposed conflict between the goals of environmental protection and economic well-being. This assumption takes the form of a hidden or buried judgment of value on the part of the analyst, which is supported—if at all—with only anecdotal evidence.

To understand the complex implications and the many levels of this all-too-common assumption about a supposed conflict between the goals of environmental protection and social economic well-being, we need to explore the popular (and unfortunate) aphorism "economy versus the environment." We shall call this the **EVE dilemma**, understanding, of course, that the word "environment" in this sense applies broadly to any system involving natural resources that can be used to support human populations.

For the last 20 years, the single premise that has most seriously hampered real progress in resolving resource-related problems in the United States (whether real or imagined, current or future) is that the interests of the environment and the economy must necessarily work in opposition to each other. The arguments are familiar and take at least two forms.

In one form, economic activity, especially production of the "hard" goods that constitute the core of the gross domestic product, requires a constant flow of energy in the form of physical resources. Constraining the extraction and processing of these resources by requiring, in whatever manner, more environmentally benign methods will raise the cost of production and hence increase inflation as a predicted result. Call this the *regulation* portion of the EVE dilemma.

In the other form, the "locking up" of wilderness, wetlands, or natural areas that might otherwise contain some development or at least supply some economic resources will inevitably shrink our resource base. Again, the likely result is more inflation, as well as the onset of an artificial condition of scarcity. People who entertain the notion of artificial scarcity have little or no acceptance of the physical reality of absolute scarcity or actually running out of resources. Call this the *preservation* portion of the EVE dilemma.

Consequently, either by excessive environmental regulation or by forced inaccessibility to the resource base, we will supposedly experience lessened availability of goods and services, increased inflation, and of course unemployment. This occurs first in certain pivotal industries and then spreads secondarily to other sectors of the economy through a reduced general power to purchase.

The three economic "horsemen"—inflation, unemployment, and decreasing production—form the grist of any scenario of economic disaster. Thus, anything that threatens to lead to any one of them, but more particularly two of them, or all of them simultaneously, must surely be anti-economic. Hence, the EVE dilemma persists.

Some people refer to the inaccessibility of a resource as artificial, implying that a sort of "natural law" guarantees humankind the right of unlimited extractive exploitation of all types of natural resources. The so-called "Sagebrush Rebellion" of a few years ago manifests this particular set of values and makes a good illustrative case.

The rhetoric of the Sagebrush Rebellion invoked the language and tone of a blatantly "free market" and the absolute "rights of private property" in discussing resources that, ironically, are clearly and obviously owned collectively by the American public. The implication was that the resources were virtually unlimited and were there for the private use of the interested parties (ranchers, miners, loggers, and so on) if only the "government" (never the "people") were to loosen its grip, which it presumably tightened out of some kind of spite and mean-spiritedness. There is, of course, an equally conventional retort on the other side of the issue.

Lost in the argument, however, are the effects on employment of the pollution-control industry and the development of technology that is *not* aimed at replacing human jobs. Although the *number* of jobs, despite appearance, is not reduced, *specific jobs* disappear; be that as it may, different jobs are created. Further, people demand more than just goods and services; they also want and gain considerable satisfaction (especially in an increasingly urbanized and affluent world) from clear air, clean water, natural areas, and so on. This sentiment was already voiced by Aldo Leopold in 1949: "We face the question whether a still higher 'standard of living' is worth its cost in things natural, wild and free. For us of the minority, the opportunity to see geese is more important than television, and the chance to find a pasqueflower is a right as inalienable as free speech."

The arguments on both sides of the economy versus the environment dilemma will not be developed further at this juncture. We have sketched out the EVE dilemma in order to develop a major premise, namely: *We will not make concrete strides toward a healthy and sustainable economic system until we rid ourselves of this unfortunate intellectual relic.*

It is, nonetheless, enlightening to summarize the historic genesis of the EVE dilemma. The debate had its origins in the United States during the expansionist 19th century "era of progress." This period, portions of which are remembered as the "robber baron era," offers an important follow-up laboratory. In this laboratory, the American experiment, supported by the immense resources of the North American continent, essentially engaged in a massive and extended test of the 18th century intellectual notion of the "Idea of Progress."

Indeed, the experiment continues in various forms, although the specter of limitations dictated by a global carrying capacity of human beings is causing considerable fraying around the edges as we look the 21st century in the eye. (Carrying capacity, as it is used here, is the number of people that can live in and use the world's environment without impairing its ability to function in an ecologically specific way.) Consider some phrases that came out of 19th century America and which form a familiar economic mythology:

- Conquering the North American continent
- Westward expansion
- Manifest destiny
- Frontier ethic
- American ingenuity
- Technological know-how

The list of these and similar "warm and fuzzy" terms could go on for some time and could be combined with institutional features, such as the Homestead Act, Railroad Act, immigration policies, and so on.

It is clear that in the 19th century the United States saw the Enlightenment idea of progress don the practical cloak of economic expansion and growth, settlement and industrialization, and ultimately urbanization. The American Industrial Revolution supported the development of the capital goods industries of the robber barons, and thus paved the way for the consumer goods economy of this century.

The legitimacy of the right to produce and consume, supported, of course, by the institutions of private property and capitalism, is therefore deeply ingrained in the American way of life. Given the flow of resources required in the unending drive to maximize the gross domestic product, the dilemma of economics versus the environment and the American way of life go hand in hand.

When it is assumed that it is one's (God-given?) duty to conquer, tame, and settle a seemingly superabundant and savage wilderness, unlimited access to resources of all kinds is seen as not only desirable but also necessary. Humans thus envision themselves as subduing and controlling the environment, not as living within it.

We must therefore seek to recast the issue of the EVE dilemma in a more productive framework. Indeed, here is the first premise to be addressed if harmony between our economy and environment is to be furthered. The source of energy for any economy comes from resources available in the environment. A healthy and sustainable economy, therefore, is dependent on a healthy and sustainable environment—*not the other way around.*

Supply-Side Technology versus Moderating Wasteful Consumption

It must be made clear up front that we are preparing to explore questions of values with respect to economics and economic society. Issues of values are always multifaceted and bring into question analytical paradigms or value systems. In order to characterize in bold shorthand the nature of the debates, which inevitably arise in resource-related issues, the following thumbnail sketches of typical competing value systems are handy:

Supply-Side Technology (Analogous to the Expansionist Economic World View[20])

- Technological optimism, which is the belief that American "know-how" will solve any problem if given a chance
- Fundamental faith in the market mechanism as the appropriate focus in the search for solutions to problems, including research
- Suspicion that involvement by the government will misallocate resources and make problems worse
- A fundamental unwillingness to accept that current levels of personal income and the quality of life are eroding

- A desire to maintain the greatest possible stability in the current economic structure and thus to expand the availability of traditional resources whenever feasible
- A ready willingness to relax the standards of environmental protection in order to treat any perceived threat to the current economic structure
- The feeling that international relations should focus on ensuring a smooth flow of resources (e.g., energy) if and when needed
- The bottom-line belief that expanding supplies is the safest and surest policy to follow given a goal of maximum socioeconomic security

Moderating Wasteful Consumption (Analogous to the Unified Systemic World View[20])

- Suspicion of technology because unquestioning faith in technology undoubtedly had something to do with causing the problem in the first place
- Skepticism about the role of the market mechanism as it pertains to environmental matters, especially in research dealing with future options and in seeking long-range sustainability
- Feeling that involvement by the government in research is necessary to ensure that currently unprofitable or unproven resources and technologies (such as wind and solar energy) are developed
- Acceptance of the notion that wasteful consumption (especially energy) has led to most of our difficulties and that some sacrifice and adjustment in our materialistic standard of living are inevitable
- Commitment to retaining and expanding efforts to protect and improve the environment
- Retaining a moral sense of intergenerational and international equity, which focuses on the issue of rich versus poor in the various stages of the WNRCP hierarchy (world to personal levels)
- The bottom-line belief that conservation of resources (i.e., ecological sustainability) is the cornerstone of rational policies toward natural resources

There we have it. Although these competing value systems are offered merely as useful characterizations, the basic framework is replicated often enough in public discourse to be instantly recognizable and usable in a shorthand analytical form. Consider a typical public meeting: Supporters of supply-side technology and supporters of moderating wasteful consumption take issue with each other over any of the various aspects of the environment versus the economy dilemma at any or all levels (world through personal) of the WNRCP hierarchy.

We hope it is clear that we are merely exploring these value systems as they affect social/environmental sustainability while neither defending nor vilifying one or the other. Our purpose in this exploration is to point out that careful and thoughtful pursuit of social/environmental sustainability can demonstrate why these supposedly conflicting value systems (around which so much social energy is wasted) are either *unnecessary* or *downright false.* Effective demonstration of this vital point will take some time and effort, and several preliminary steps must be taken first, such as identifying our world view.

IDENTIFYING OUR WORLD VIEW

One of the mistakes economics has historically made is a failure to identify assumptions concerning human values inherent in the world view being employed at the time. As we shall continually try to demonstrate, there is no intellectual thought that we know of which is free of judgment based on human values. In other words, a human being cannot hold a value-neutral thought in his or her mind because we are by nature subjective beings and *cannot* be truly objective, much as we might like to be.

We attempt to avoid making this same mistake by identifying a clear set of premises that we are making about the world. We do this through a simple identification of assumptions based primarily on scientific data as we understand it and then on a two-part exercise in logic.

Two-Part Exercise in Logic

To begin with, we assume that the carrying capacities of the various resource systems of Planet Earth are under considerable stress, which is likely to get worse in the immediate future. Attempts to identify systems that offer evidence of this premise might include items on the following list, which, although imposing, is nonetheless pitifully inadequate:

- Loss of soil to erosion and declining fertility
- High levels of pollution and declining production in the oceans

- Depletion of the ozone layer and its shielding capability
- Loss of species, biodiversity, and genetic diversity
- Worldwide deforestation and desertification
- Impending threat to traditional supplies of energy and the environmental consequences of converting increasingly to such things as nuclear energy
- Many local, regional, national, and global problems related to the burgeoning human population as well as its tremendous mobility

This purposely incomplete list is virtually random and could go on for some time, but that is not the point. Our purpose is to make informed decisions about the evidence presented by others and to propose ways of behaving accordingly, *not* to provide proof with respect to these or similar impacts the world may be experiencing. Above all, our work is about doing something, not just analyzing something. Having said this, we hasten to add that each individual's assumptions about the state of the world are correct from his or her point of view and, as such, are critical and must be made clear.

Of course, information on the health of the world and its local and regional support systems is vitally important because each person ultimately has the duty to personally evaluate such information. Many excellent organizations and individual scientists (biological, physical, and social) have compiled exhaustive data and continue to collect and monitor these well-publicized effects. So don't take our word for it. There are much better sources. To cement the main point of our analysis, however, consider the first step in logic.

Logical Step #1

There are three ways of seeing alleged impacts and their effects on our world. They can be seen as *true, false,* or *inconclusive*. In the third case, there is not enough perceived evidence to say one way or the other, and some people would argue that everything is this way and always will be. Be that as it may, each person must choose one of the above options as a basis for day-to-day behavior, and such choice is either overtly or indirectly acted out through one's behavior.

In a very narrow sense, it would seem possible that data regarding some alleged impacts are true and others false. Given the well-documented interactive and interdependent nature of the various resource systems of the globe, it is unlikely that, in a broad and workable sense, some data are true and others false. If some are true, the Earth is destined to suffer an impairment in its carrying capacity, and the exact details of the scenario may be relatively unimportant. One failed system may apply pressure to and threaten the collapse of another, which until that point may have appeared stable.

For instance, assume that an apparently healthy land-based agricultural system produces sufficient food for the world's people, but only if accompanied by an adequate nutritional contribution from the oceans. If the oceans suffer a collapse, however, the land based agriculture may prove inadequate, given the number of mouths to feed. This could be true in the sense of simple aggregate production, even if the land and oceans were not ecosystemically interconnected and interactive—which of course they are.

The conclusion of the logical first step addresses which of these we are to assume (true, inconclusive, or false) and why. If one assumes either true or false, the operating principles one will support are clear. A selection of true means that one pursues the principles, concepts, and goals of sustainability. Conversely, a selection of false means that one will see merit in promoting a policy of full speed ahead within the framework of our current economic system.

There are interesting implications surrounding the "middle ground" when one assumes the evidence for impairment of the world's carrying capacity is inconclusive. Uncertainty about the future state of the Earth's support systems is, of course, the position occupied by a large number of people. Reasons vary but include not having seriously considered the question, accessing apparently conflicting data, and values and vested interests that lead to skepticism that human society faces limits. At least such is often the case from a personal, community, or bioregional point of view. Nevertheless, if one assumes "inconclusive," one must still act as if the data were true or false because one cannot act from inconclusiveness.

There are many fascinating directions we could take in playing off this notion, including:

- The tendency not to consider environmental issues in depth because the issues are "depressing" or because one is afraid of what one will learn
- The tendency to believe only information that one wants to believe because of special interests
- The tendency to concern oneself with everyday "getting ahead" and to say, in effect, "I'll become an environmentalist *after* I'm wealthy"

- The tendency to conclude that any stress to the global carrying capacity will fall either on resource systems that will not affect one's personal life or in communities, regions, or countries other than one's own
- Of course, there is always the almost blindly optimistic devotee of the American Dream who believes that technology will provide a way, almost regardless of what the evidence or data might say

Pursuing any of these at this point would divert us from the task of concisely finishing the chosen line of thinking. What do we think? Our position hinges on two points.

First, it is difficult to see how anyone could seriously examine the continually mounting evidence (biological, geophysical, medical, chemical, climatological, and so on) and end up with an honest selection of false, concluding that there are definitely no problems on the horizon. Since the role of technology has not been explored, however, one could conceivably choose the inconclusive category.

Second, we must remind ourselves that the most important "real-world" question facing humanity today is hanging in the balance: How long will our home planet be able to support human life as we know it? This question is important because a mistake that negatively affects the Earth's sustainability will be very costly.

Consequently, because of the extremely high stakes involved, prudence would dictate that one act in the same manner as if one had selected either true or inconclusive with a leaning toward true. After all, if sustainability is consciously promoted and it turns out that we have more carrying capacity than we thought, the result would be a pleasant surprise for our descendants instead of a nasty shock.

The result, in our minds at least, is a modern-day Pascal's Wager, whereby if we assume sustainability to be the only viable path, we and our children and their children stand to either gain everything or lose nothing. If, on the other hand, we continue operating as though it is a world without limits, we and all future generations stand to either gain nothing or lose everything. To us, the choice is clear.

Logical Step #2

Our final logical step can be concisely summarized. Once we set our tent firmly in the "true" camp, we begin the quest of uncovering the operating principles of sustainability. But where do we begin? How do we proceed?

It appears that the single most important act of modern society that threatens the global carrying capacity is *production* (or *extraction*) of economic goods and services from the biosphere. Expansion of production in the aggregate is *growth*. We must therefore reconsider that most hallowed element of modern industrial society—economic growth, which means that we must account accurately for its costs, as well as its highly touted benefits, at all stages (from the world level to the personal) of the WNRCP hierarchy.

Here it must be pointed out that many people focus on a highly visible problem that seems, for the moment at least, a more obvious cause of the world's strained carrying capacity, and that is the burgeoning human population. It is also economically safer to conclude that any problems with the carrying capacity of the world hinge on too many people, *not* on too much production of economic goods and services. So the question becomes: What is the primary cause of the world's social/ecological woes with respect to observed stresses of its carrying capacity—human population or economic production?

While the number of humans the globe is being asked to support is obviously a monumental issue, we conclude that the direct pressure on resources is created not only by efforts to increasingly extract economic products to support those people but also by increasingly intense advertising to foster more and more consumerism, which in turn creates demand for more economic production. Therefore, the focus of analytical attention properly belongs on *economic production*.

In summary, the mix of apparent fact and choice of values identifies our "world view." It can be stated as follows:

1. Evidence appears overwhelming that many resource systems either are or will soon be facing stresses to their carrying capacity of crisis proportions.
2. The self-reinforcing feedback loops that compose all of Nature's support systems are interconnected and interdependent; abuse of them constitutes a generalized threat to the carrying capacity of our home planet.
3. Despite the clear role of growth in human populations, economic production (as a specific human act) is the main culprit in directly causing the apparent stress of the world's carrying capacity.
4. Since *some* level of economic production could be sustainable for centuries, the initial focus of attention must be on changes in that level of aggregate production. This, of course, amounts to *economic growth*.

With our world view in mind, it is time to consider economic growth from the planetary scale to the personal.

ECONOMIC GROWTH FROM THE PLANETARY SCALE TO THE PERSONAL

Before reviewing the ideas of modern writers, it is useful to identify what conventional economics has, from its inception, thought it was up to. This is particularly necessary if we are to address the growth ethic from a point of human values. A brief journey to the origins of modern post-Renaissance thought seems in order.

Historical Perspective

To begin, economics is first and foremost a social science. This means that the behavior of people is inherent in its assumptions and results. For better or worse, the beginnings of economics as a discipline supposedly occurred with Adam Smith and the publication of *Wealth of Nations* in 1776. (People who are adamantly growth oriented note that the birth of economics is coincident with the beginnings of the nation destined to provide the most aggressive testing ground for these classical liberal ideas.)

It is significant that Adam Smith was every inch a product of the Enlightenment, which, over a period of three to five centuries, ushered in an extended love affair with the notion that human analytical thought can, among other things, uncover absolute truth. This culminates a process marked by shifts in the dominant view of Western society from reliance on faith to reliance on reason, from divine revelation to empirical scientific discovery, and from the concentration of human affairs within the realm of the transcendental and spiritual to the realm of the secular and material.

From the Middle Age doctrine of the Divine Right of Kings, Western society experimented with the notion that it could discover physical principles of the world through observation and reason. It also experimented with the notion that it could structure its own economic and political institutions to be responsive to the democratically expressed will of the people, that institutions could be made to serve the people, rather than the reverse.

But we digress. Thousands of volumes have been written on all aspects of these powerful ideas, and it is not our purpose to recreate the major tenets of intellectual history. The important point is that, as a typical product of the times, economic methodology became enamored from its outset with empirical scientific method and thus preoccupied with the self-appointed task of becoming a science.

Social philosophers of the 18th century Enlightenment were so bedazzled by the astounding 17th century scientific discoveries in astronomy, physics, mathematics, plate tectonic geology, chemistry, and so on that many set off in a fervent search for the universal "laws of motion" of human behavior (the "nature" of human nature, if you will). Certainly, there must be deterministic laws for human behavior that parallel Isaac Newton's law of gravity. As a species, we have perhaps never been able to avoid spending too much time playing with a novel new toy.

If, therefore, the new body of analytical method was to be able to call itself a science, it must somehow inject absolute predictability into the material, acquisitive actions of human beings. After all, an apple *always* falls from a tree in the same manner. The result is a bundle of assumptions known as Rational Economic Man.

Major characteristics of Rational Economic Man are:

- Self-interest (self-centered)
- Complete and perfect knowledge of all alternatives (one thinks)
- Acquisitive
- Materialistic
- Believing more is better—always preferable to less
- Preferring immediacy—something now is preferable to something later
- Always making the same choices ("rational")
- Competitive in behavior

Additional characteristics of Rational Economic Man are explored in the next section, along with several other common terms and concepts from the separate but connected worlds of economic theory and practice. Those selected are chosen because of their particular usefulness in exploring the full implications of the preoccupation of our society with the phenomenon of economic growth as part of its philosophical underpinnings. Suffice it to say, Rational Economic Man stands as the definitive theory of human nature emanating from the classical beginnings of the discipline of economics.

Growth/No-Growth Tug-of-War

Given that we have accomplished the preliminary steps of identifying our world view and developing some necessary methodological background, we turn to the first question: What are the important features of the relationship between sustainability and traditional economics?

Careful thought about the issue of achieving sustainability at any particular level (world to personal) of the WNRCP hierarchy reveals a fundamental uneasiness between the notion of sustainability and growth-oriented economic theory and practice, which can be likened to a growth/no-growth philosophical tug-of-war.

Clearly, the growth/no-growth tug-of-war is similar to the economic versus environment (EVE) dilemma, but is even more fundamental and pervasive. The EVE dilemma merely describes practical working behavioral assumptions of various participants on the stage of day-to-day actions. These are the assumptions espoused by businesspeople, public officials, environmentalists, and so on. It is what one hears in public forums on one side of the aisle or the other.

The growth/no-growth tug-of-war, on the other hand, is the more fundamental methodological and philosophical conflict between the goals of sustainability and economic theory. But is this conflict real or perceived? At what levels (world to people) of the WNRCP hierarchy does it apply, if at all? Can economic theory and methodology be harnessed productively to work for the benefit of the task of building sustainable communities?

If sustainability is to survive in the face of practical economics, there can be no more important task than sorting out the philosophical conflicts implied by the growth/no-growth tug-of-war, one of which is scarcity.

Scarcity

Perceived scarcity and the fear of loss or of not having enough to fill our necessities are born out of a perceived threat to our "right of survival," however that is defined. The perceived security of our right to survive is weighed against the number of choices we think are available to us as individuals and our ability to control our choices. Perceived choices are ultimately affected by the real supply and demand for natural resources, the source of energy required by all life in one form or another.

The greater the supply of a particular resource, the greater the freedom of choice an individual has with respect to that resource. Conversely, the smaller the supply, the narrower the range of choices unless, of course, we steal choices from other people to augment our own. And scarcity, real or perceived, is the breeding ground of both environmental and social injustice, which rear their ugly heads each time someone steals from another rather than taking responsibility for his or her own behavior and sharing equally.

Overexploitation of Resources

According to a song popular some years ago, "freedom's just another word for nothing left to lose," which in a peculiar way speaks of an apparent human truth. When one is unconscious of a material value, one is free of its psychological grip. However, the instant one perceives a material value and anticipates possible material gain, one also perceives the psychological pain of potential loss.

The larger and more immediate the prospects for material gain, the greater the political power used to ensure and expedite exploitation, because not to exploit is perceived as losing an opportunity to someone else. And it is this notion of loss that one fights so hard to avoid. In this sense, it is more appropriate to think of resources as managing humans than of humans as managing resources.

Historically, then, any newly identified resource is inevitably overexploited, often to the point of collapse or extinction. Its overexploitation is based, first, on the perceived rights or entitlement of the discoverer/exploiter to get his or her share before someone else does and, second, on the right or entitlement to protect one's economic investment. There is more to it than this, however, because the concept of a healthy capitalistic system is one that is ever growing, ever expanding, but such a system is not biologically sustainable. With renewable natural resources, such nonsustainable exploitation has a "ratchet effect," where to ratchet means to constantly, albeit unevenly, increase the rate of exploitation of a resource.[21]

The ratchet effect works as follows: During periods of relative economic stability, the rate of harvest of a given renewable resource (timber or salmon, for example) tends to stabilize at a level that economic theory predicts can be sustained through some scale of time. Such levels, however, are almost always excessive, because economists take existing unknown and unpredictable ecological variables and convert them, in theory at least, into known and predictable economic constants in order to better calculate the expected return on a given investment from a sustained harvest.

Then comes a sequence of good years in the market, or in the availability of the resource, or both, and additional capital investments are encouraged in harvesting and processing because competitive economic growth is the root of capitalism. When conditions return to normal or even below normal, however, the industry, having overinvested, appeals to the government for help because substantial economic capital, and often many jobs, is at stake. The government typically responds with direct or indirect subsidies, which only encourages continual overharvesting.

The ratchet effect is thus caused by unrestrained economic investment to increase short-term yields in good times and strong opposition to losing those yields in bad times. This opposition to losing yield means there is great resistance to using a resource in a biologically sustainable manner because there is no predictability in yield and no guarantee of yield increases in the foreseeable future. In addition, our linear economic models of ever-increasing yield are built on the assumption that we can in fact have an economically sustained yield. This contrived concept fails, however, in the face of the biological sustainability of the yield.

Then, because there is no mechanism in our linear economic models of ever-increasing yield that allows for the uncertainties of ecological cycles and variability or for the inevitable decreases in yield during bad times, the long-term outcome is a heavily subsidized industry—corporate welfare, if you will. Such an industry continually overharvests the resource on an artificially created, sustained-yield basis that is not biologically sustainable.

Because the availability of choices dictates the amount of control one feels one has with respect to one's sense of security, a potential loss of money is the breeding ground for environmental and social injustice. In this kind of injustice, the present generation steals from all future generations by overexploiting a resource—thus creating scarcity in both the resource itself and in the choices pertaining thereto—rather than giving up potential income. In addition to the scarcity caused by outright overexploitation, the term "scarcity" has another facet, one exemplified by the Oil Shock, which requires some background to understand.

Oil Shock and the Future

In October 1973, the Oil Producing and Exporting Countries (OPEC) ushered in an oil embargo that became known as the Oil Shock. It was marked primarily by rapidly rising prices of crude oil. Until October 1973, the price of crude oil had been remarkably stable, holding between two and three dollars per barrel for more than ten years (1960 into the early 1970s). Beginning in late 1973, the remainder of the decade saw fitful increases, which reached as high as $34 per barrel in the early 1980s.

Since oil provided 40 percent of the world's energy needs, the resultant shock waves to the world economy were severe. But why not view it simply as a "price shock," albeit to an important resource, but one that would nevertheless be overcome in due time with the involvement of technology?

The answer lies in the fundamental pervasiveness of energy as a required component of virtually all economic processes. Additionally, at the outset of the Oil Shock, the permeating effects of energy (and fossil fuels in particular) were not understood as well as they might have been, even by professionals whose duty it was to know better. An anecdote will best serve to underscore this point.

In the heat of the initial inquiry into the economic effects of the Oil Shock, and during the height of the "stagflation" (a condition in which a high rate of price and wage inflation is coupled with stagnant consumer demand and high unemployment) experienced in 1974, economists attempted to estimate the proportion of the double-digit inflation that was based on energy. Since empirical evidence suggested that direct purchases of energy averaged 8 percent (or about one-twelfth) of the total costs or expenditures by U.S. firms, that figure was put forward. After all, the other 92 percent (nonenergy expenditures) would in effect tend to shield a producer from energy-related inflation.

Be that as it may, inflation rose from 4 percent in 1971 to as high as 12 percent in 1974. The above method of explaining this increased inflation suggested that only about two-thirds of one percentage point could be ascribed to energy, which is not much when one considers that inflation still rose 11.3 percent without the effect of energy prices.

Subsequently, however, the tool of analyzing the input versus the output was employed to reestimate these figures. This was done because an input/output analysis had the ability to recognize cumulative economic effects by measuring the flow of materials. The input/output analysis begins with raw materials from primary suppliers and follows the flow to the intermediate producers of goods and on to the final producers of goods and services in whatever manner is determined by the underlying structure of the economy.

In other words, an input/output analysis would recognize that the purchase of a component for a larger product (such as a tire for an automobile) would represent the purchase of previously used energy (which is stored up in the purchased tire). The logic is much like that underlying the original labor theory of value as developed by Adam Smith, David Ricardo, and Thomas Malthus. According to the classical school of economics, the source of all value is labor because it is indispensable in the production of all goods and is thus inherently represented in the purchase of all goods and services.

Using the techniques of an input/output analysis, which reflects the many stages of complexity in the U.S. economy, the new and dramatically increased estimate of inflation indicated that about two-thirds of it was accounted for by the energy sector. This means that the rate of inflation would have increased from 4 percent to between 6 and 7 percent in the absence of the Oil Shock, as opposed to 12 percent with the Oil Shock. The rate of inflation without the Oil Shock would seem plausible, given that economists were ascribing some inflation to the delayed effects of deficit spending during the late 1960s in order to wage war simultaneously on poverty and North Vietnam.

The jump in the estimate of inflation—from less than 10 percent to over 60 percent—just because of using a different method of analysis suggests that economists did not correctly understand the role of energy in our overall economic structure. Energy per se had been ignored and thus treated like any other input.

The analysis was standard; if the price of a commodity were to increase, the users of that commodity would simply substitute something else for it. Unfortunately, there is no substitute for energy, and because it is ubiquitous in all productive processes, a temporary situation simulating absolute scarcity is set up, for which inflation is the only safety valve.

As a side comment, it was easy to take for granted the role of energy, since the price of gasoline had been approximately 35 cents per gallon just before the Oil Shock hit in 1973, although it had already exceeded 30 cents per gallon as early as the late 1940s. This means that, supported by the continual realization of economies of scale in production and refining, the nominal price of gasoline had remained stable for about a quarter of a century. (Economies of scale is the cheapening of production as the size of a company increases.)

This amazing fact, coupled with the moderate level of price increases in gasoline during that period, meant that for 25 years the *real* cost of energy declined steadily relative to all other inputs. Entrepreneurs, as well as consumers, did exactly what economic theory would predict; they substituted in the direction of the relatively less expensive input—and the American economy went on a monumental binge of energy consumption.

In addition to the simple glut in the consumption of energy, a tremendous amount of technological change took place during that period. Innovation, which could afford to take energy for granted, was focused in an energy-intensive direction and invariably demanded a greater expenditure of energy per unit of output than did previous processes.

Further, since the increased use in energy and the adoption of new capital-based technology became almost synonymous, the already low and falling real price of energy stimulated many capital/labor substitutions. Indeed, the search for new ways to mechanize our daily lives remains a favorite pastime of American business.

A rosy picture was, of course, painted for the employed labor force, because the increasing availability of capital led to empirically measurable increases in productivity, or output per worker. Thus, even in the face of high rates of economic growth, unemployment began to be a nagging problem—and now the term "scarcity" was beginning to command center stage.

To understand the context of scarcity, one must look beyond the energy sector. The era of environmental awareness had definitely begun by 1973. The first Earth Day had taken place in 1970, and the original study on *Limits to Growth* had been published in 1972.[22] In many other ways, activists, scholars, public officials, and private citizens were beginning to respond to the issues of environmental health and the potential depletion of resources.

Regardless of the details of how the Oil Shock occurred—or perhaps was engineered—the message from it and other environmentally related phenomena to the industrialized world should have been clear: It is more and more likely that society will face a future in which many natural resources (especially the "marquee" resource, petroleum) will become increasingly short in supply. Although technology can move the problem around by shifting the stress of scarcity from this resource to that, it cannot simultaneously relieve the pressure of scarcity on all resources. Consider the following vignettes:

1. If oil is in short supply, we could use more coal, but that causes acid rain, puts pressure on supplies of water and clean air, and affects forest health and global warming.
2. If supplies of food are questionable, we could bring more land into production and farm more intensively, with a higher degree of dependence on chemicals, but that puts pressure on forests and other nonagricultural lands, on water and fossil fuels used for fertilizers and pesticides, and so on.
3. If acid rain due to combustion of coal and other fossil fuels threatens lakes and forests, we could seek to develop new and cleaner technologies and/or introduce new and cleaner sources of energy on a massive scale.
4. If population pressures lead to urbanization and the conversion of productive farm land to urban uses (as is now occurring), then return to #2...

The possibility of additional relevant vignettes is endless, and we invite you to write your own. However, some significant general principles emerge:

- The use of energy is inherent in all we do—both the initial creation of a problem and the search for and implementation of its solutions.
- Technology, even if strikingly successful, merely shifts pressure from one resource system to others. We can only hope that the newly pressured systems have considerably more carrying capacity than the old.
- The role of time is critical in two respects. First, the lead time in technological development must be short enough to avoid a serious crisis. Second, we are merely buying time with technological innovation. How long will the technological fix last?
- Scarcity applies not only to such resources as petroleum and available land for farming but also to such resources as clean air, clean water, healthy soils, a healthy layer of ozone, and a climate that is free of human-induced warming.
- Specifically, the quality of our air plays a unique role because polluted air pollutes everything from the top of the highest mountain down into the soil, where it is carried by rain and melting snow. Hidden in the soil, water collects and flows down-

> hill to emerge in streams and rivers on its journey to the sea. The sea, having no outlet, concentrates the pollution it continually receives down into its very depths. The quality of the world's air may therefore prove to be the ultimate limiting factor in the survival of human society.

Thus, the Oil Shock and the energy crisis were merely the tip of the iceberg in ushering in the era of concern about the perpetual, generalized, and growing scarcity in resources. Due to the widely held American myth that technology is our savior, we, in blind faith, expect technology to find a way to, as in the "Perils of Pauline," snatch the fair maiden from the tracks in front of the onrushing Scarcity Express at the last second.

Many of our "quick fixes" are therefore publicly viewed and touted as technological triumphs, rather than being seen as band-aid solutions that shift the stress to other resources and even set up more serious problems some time in the near future. Significantly, the stress, as the vignettes suggest, is often shifted from visible supplies of some resources to less visible environmental factors, including air and global climate.

The issue of the Oil Shock resulted almost instantly in public awareness of the involvement of energy with most other physical resources as well as general environmental variables, all fueled by the reality and mythology of technological innovation. To us, events of the last 20 to 25 years suggest that attention paid to this collection of factors represents no less than an alternative world view or way of viewing the future, which encompasses the entire general phenomenon within the term *scarcity*.

There are important lessons in all of this for those involved in sustainable community development. First, history suggests that a biologically sustainable use of any resource has never been achieved without first overexploiting it, despite historical warnings and contemporary data. If history is correct, resource problems are not environmental problems but rather human ones that we have created many times, in many places, under a wide variety of social, political, and economic systems.

Second, the fundamental issues involving resources, the environment, and people are complex and process driven. An integrated knowledge of multiple disciplines is required to understand them. These underlying complexities of the physical and biological systems preclude a simplistic approach to sustainable community development. In addition, the wide natural variability and the compounding, cumulative influence of continual human activity mask the results of overexploitation until it is severe and often irreversible.

Third, as long as the uncertainty of continual change is considered a condition to be avoided, nothing will be resolved. However, once the uncertainty of change is accepted as an inevitable, open-ended, creative process of life, most decision making is simply common sense and must take into account Herman Melville's admonishment: "Our actions run as causes, and they come back to us as effects."

For example, common sense dictates that one would favor actions that have the greatest potential for reversibility, as opposed to those with little or none. Such reversibility can be ascertained by monitoring results and modifying actions and policies accordingly, but to understand this, we need to consider more closely the language of economics by examining some terms.

6

THE LANGUAGE OF COMMON ECONOMIC CONCEPTS

This chapter plays the dual role of exploring some all-important economic concepts and providing a necessary transition to applying the notion of sustainability to both ecology and economics. The framework within which we will discuss economics was developed in Chapter 5. It identified our point of view and assumptions about values and concluded with a careful assessment of the phenomenon of scarcity, which must become the centerpiece of economic thinking and methodology if we are to harness economic growth in the interest of achieving social/environmental sustainability.

Chapter 7 will begin the job of forging and nurturing a practical working relationship between economics and social/environmental sustainability, which is a major goal of this book. As a prelude, however, careful consideration must first be given to commonly used terminology. Clarity of language is all-important if productive and concrete steps are to be taken, because "economics" and "sustainability" do not appear to be comfortable bedfellows.

After all, the historical promotion of growth has occurred not only because economists have applied their formal disciplinary tools but also because the many and diverse people dealing with business in both the public and private sectors have been equally ardent in championing unbridled economic growth. The attempts to harness economic jargon in the service of a particular vested interest or point of view have been nothing short of monumental. The purpose of this chapter is to sort out some of the resultant terminology.

As mentioned in Chapter 5, we have done our best to humanize the discussion of economics because it has developed into a very depersonalized discipline. Despite our best attempts, however, the chapters on economics—especially this chapter—may appear to be written in a different style than the rest of the book.

THE GENERAL ROLE OF SEMANTICS

Economic terminology is rampant with words, phrases, and concepts that are used and misused in both criticizing and defending our beliefs and habits. Certainly, misunderstanding important economic terms is dangerous, if for no other reason than it blocks effective communication. Such misinterpretation will lead people to think they understand one another (when they do not) and have therefore reached a possible agreement (when in fact they have not).

By identifying five critical economic concepts, we will attempt to wade through the necessary jargon as productively and painlessly as possible. We will discuss the following for each concept: (1) how the term has been used professionally by economists within the discipline of economics; (2) how the term is used in practice, including the way in which everyday usage might innately frustrate the implementation of attitudes and policies that potentially support social/environmental sustainability; and (3) suggestions for alternative interpretations of each term, which will be solidified in the next chapter.

In short, we explore exactly what each term is supposed to mean and how it has been used to promote economic growth, from theory to practice. In so doing, we set the stage for redirecting economic methodology and the practical application of economic thought along the lines of sustainability.

The Tension Between Theory and Practice

It goes without saying for any discipline that exploring differences between commonly accepted usage of terminology and alternative, more desirable usage, which some might call utopian, is tricky business. Would you tell a physicist that he or she does not understand the difference between the concepts of energy and power or a biologist that he or she has missed the essence of an ecosystem?

Clearly, no one is going to overthrow and rule as obsolete the discipline of economics with its long history of applicability in "practical" affairs, and that is not our goal. We find it far more constructive, if it is possible, to simply adapt terminology to the task of meeting real human and planetary necessities as they are identified.

Since our view is that the *real* long-term necessity is sustainability, the general methodological task before us is well defined. For too long, there has been a tendency for people to serve economic dogma. We seek to reverse this trend and have economics serve people and the environment—from which both people and economics are inseparable.

It is probably an overly ambitious goal to humanize the use of economic analysis, but if the shoe fits.... Our main premise throughout this book remains unabashedly optimistic: a broad, comprehensive, value-based consideration of economic philosophy and methodology can lead to a rich and rewarding marriage with social/environmental sustainability, *not* that sustainability and economic theory and practice are incompatible.

This marriage cannot be consummated, however, through the typical approach, which is to persistently seek narrow technical definitions and uses of all terms.

It must be noted that this legacy is not confined just to noneconomists misusing terms about which economists think they know better. Economists are occasionally guilty of injecting value-laden prescriptions or opinion disguised as value-free analysis through the subtle use of economic terminology. Thus, each camp, the professional and the practical, has its abusers.

Our critique is thus directed as much at the formal discipline of economics as it is at businesspeople and public officials who uncritically adopt as valid a poorly understood piece of economic theory just because it fits their preconceived biases and vested interests. In truth, all analysis is value based, if for no other reason than we humans are by nature subjective creatures and cannot be otherwise, no matter how hard we may try. For communication to be accurate, therefore, it demands the up-front acknowledgment and precise identification of the role played by a particular set of attitudes and values.

"But," you might ask, "why worry about theory at all when its sterility and lack of applicability to the so-called 'real world' have long been noted by the practical, results-oriented movers and shakers of our society?" True, we've all heard (and probably said), "That's fine in *theory,* but it will never work in practice." This derisive statement is, of course, meant to irrevocably relegate to the conceptual slag heap some idea with which we disagree.

The problem is that all practical behavior is rooted, wittingly or unwittingly, in some theoretical construct. *Homo sapiens* is a creature of subjective abstract thought, even those of us who think of ourselves as "practical people." The mere ability to recognize patterns in everyday life and act predictably and accordingly offers each of us continuous practical applications of theory.

For instance, if you back your automobile out of your driveway and head down the road, you automatically drive on the right side of the road (in the United States at least), as will the oncoming vehicles. This is a simple example of social order, which is designed to, and usually does, avoid the chaos of head-on collisions. There are no signs telling you to do this, but previous behavioral patterns and images connect in your brain, and you act accordingly. Hundreds of routine daily actions offer similar examples.

Many people who are frustrated with the pace and direction of contemporary society will quickly agree with the contention that the current economic paradigm overly dominates the fabric of modern life. It makes little sense, however, to launch into a detailed prescriptive analysis, particularly with a goal of developing sustainability-related operating principles, without first forging a better common understanding of the language employed. In this sense, our critique must extend to both theory and practice and in particular must emphasize the often nebulous connection between the two.

Setting the Agenda

We have chosen five well-known and commonly used terms. The list is not comprehensive; others could have been chosen. But, as will become apparent, one term suggests another, and so on. There is no unique list of concepts that must be scrutinized in order to define effectively the differences between the typical use of economic terminology and a use supportive of social/environmental sustainability. The five concepts are (1) Rational Economic Man, (2) self-interest, (3) the invisible hand, (4) productivity, and (5) technology.

Given the importance of clarity at this juncture, our approach is systematic, careful, and even repetitive. For each of the above concepts, the discussion will proceed within three headings: (1) the term and economic theory, (2) the term in practice, and (3) the term and sustainability.

Through this approach, we seek to achieve the following: (1) to uncover common threads or features with respect to either the current or desired usage of a term within the general principles of economic theory, (2) to reveal the heart of the relationship between economic theory and economic society, and (3) to develop alternative premises to such concepts as Rational Economic Man when defining the economic behavior of human beings.

Finally, this new understanding can be used to systematize, strengthen, or streamline the applicability of that theory to the problem at hand—the practical implementation of community sustainability within a sustainable landscape. The underlying human behavioral attitudes and traits, whether desirable or undesirable, can be exposed and integrated into the analysis.

Remember that economics is by definition a social science and necessarily operates on a foundation of assumed behavioral motivations and traits of people. As of yet, we are stuck with the concept of Rational Economic Man as a synthesis of human behavior, which offers an excellent place to begin examining the terminology.

THE TERMS

Rational Economic Man

Rational Economic Man and Economic Theory

As the dominant behavioral paradigm permeating conventional economic theory, Rational Economic Man is the ideal starting point for our consideration of terminology. Recall that the earlier bundle of assumptions we entitled "Rational Economic Man" were identified as the summary outcome of a statement on human nature by the philosophers of the Enlightenment, who were responsible for early economic thought. The features assumed by economists to define Rational Economic Man are repeated here:

- Self-interest
- Complete and perfect knowledge of all alternatives
- Acquisitive
- Materialistic
- Believing more is better—always preferable to less
- Preferring immediacy—something now is preferable to something later
- Always making the same choices ("rational")
- Competitive

Economists employ various statements and levels of intensity in defending the use of these assumptions. The most pragmatic defense is that *they make the model work.* This is not surprising, however, because people of their own volition act primarily in accordance with the assumptions, and economists note that exceptions are not pronounced enough to render invalid the package called Rational Economic Man. And, of course, the "industrial strength" defense is that's just the way people are. With this in mind, it is useful to ask the reasons *why* this image is not only so pervasive but also preserved and perpetuated.

To the uninitiated eye, the answer would appear to be that the traits accurately describe human nature and behavior. Anyone who builds mathematical models realizes that realistic assumptions generate useful models because they have the ability to accurately predict real-world phenomena. Economists, among others, go to great lengths and invoke elaborate semantic gymnastics to convince themselves and others of the "fact" that people really do act this way.

It is thus instructive to turn this whole notion of self-interest and rational behavior back on economists themselves. In so doing, we must somewhat cynically contend that a major reason professional economists cling to the spartan assumptions of Rational Economic Man is to make their economic theories and models work.

The procedures with which to analyze the optimization and equilibrium required by most theoretical models simply will not work without that appropriate "something" to optimize—the assumptions of Rational Economic Man. Further, that "something" must ideally be tangible, consistent, and quantitatively measurable if at all possible. More often than not, that "something" turns out to be human satisfaction as represented by acquisition of wealth in the form of money and consumer goods.

To allow Rational Economic Man to become capricious, flighty, indecisive, and unpredictable would destroy an economist's personal investment over an entire career in mastering an esoteric and elegant body of method. Self-interest suggests that this would be viewed as the ultimate threat to legitimacy, that is, to the relevance of the analytical tools themselves.

Although no one wants to believe that what he or she has spent a career learning—and thinks he or she *knows*—is suddenly obsolete (and must therefore be defended at any cost), we believe this situation has been responsible for many an unfortunate historical circumstance. Nevertheless, in practice, human goals and behavior do not always square with the assumptions of Rational Economic Man, which should a priori serve to call into question the "universal" predictability and usefulness of any model purporting to rely on it.

When confronted with evidence of human behavior that is inconsistent with the assumptions of Rational Economic Man, the instinct of self-preservation often leads economists to point out that a particular assumption was not *really* granted or that a twist in language, which would indicate an apparently altruistic behavior, was *really* self-interest, and so on. If nothing else, they retreat to the premise that *most* people act consistently with the assumptions of Rational Economic Man *most* of the time, and that should be sufficient to ensure the validity of the theory.

Of course, if exceptions do exist, monstrous errors in describing the goals and behavior of people can occur, as pointed out by Sherlock Holmes to Dr. Watson: "While the individual man is an insolvable puzzle, in the aggregate, he becomes a mathematical certainty." Holmes goes on to explain that it is impossible to foretell what any individual person will do, but an average number of people are always predictable because, although individuals vary, percentages remain constant.

Sherlock Holmes, in his discussion of the predictability of human behavior, touches the core of special cases and common denominators. He saw each person as a special case and therefore unpredictable, but if we study enough special cases with an eye for their common traits (common denominators), then we can make certain predictions about their generalized behavior. Swiss psychiatrist Carl Jung put it differently.

Jung asserted that self-knowledge is a matter of knowing individual facts, so theories are of little help. The more a theory claims universal validity, he said, the less it does justice to individual facts. Any theory based on experience is thus necessarily *statistical*, which means it formulates an ideal average and abolishes all exceptions by replacing them with an abstract mean. Although the mean is valid and is an unassailable, fundamental fact in the theory, it need not occur in reality. In addition, the exceptions of either extreme, though equally factual, do not appear in the final result because they cancel each other out.[23]

If, for example, one were to weigh each pebble in a jar of pebbles and get an average weight of 145 grams, Jung said one would still know very little about the real nature of the pebbles. Anyone who thought, on the basis of this average, that he or she could pick up a pebble weighing 145 grams on the first try would be disappointed. Indeed, however long one searched, one might never find a pebble weighing exactly 145 grams.[23]

Despite the obvious validity of the assertions by Sherlock Holmes and Carl Jung, so much effort has been invested in developing and employing the theory of Rational Economic Man that we would almost rather force some economic actor into an uncomfortable and poorly fitting box than admit irrelevance of the theory or model. Any economist knows (and in nonthreatening circumstances is quick to admit) that a theory is only as good as its assumptions, and if those cannot be granted, the resulting model may be little more than an amusing mathematical toy, unable to predict real human behavior.

Rational Economic Man in Practice

Do people really behave according to the assumptions of Rational Economic Man? Countless analysts, businesspeople, and corporations fervently hope so. The person who does fit the assumptions is a very compliant worker and consumer, given the sparseness of their philosophical framework and the materialistic nature by which they are assumed to be motivated.

Undeniably, human beings invest much time and energy in what can be called "maximizing behavior." We try to secure the best-paying job we can, and once secured, we try to spend our income in the most prudent manner consistent with our beliefs and tastes. Surely there is nothing in this familiar effort to maximize income and spend it satisfactorily that would make the economic theory of demand not work.

To Rational Economic Man, however, additional acquisition is always the order of the day. We may become satiated with a particular product, but a generalized desire for more of other things continues unabated. If, for instance, one gets enough hamburger, one looks to steak. If one gets enough food, one concentrates on housing, clothing, entertainment, and so on. As the philosopher Walter Weiskopf has written: "When your credo is 'More is Better,' then Enough never comes."

The behavior elicited by the credo "more is better" can seem particularly frantic in a complex, affluent society, such as ours, which has (according to a recent count) between 30,000 and 35,000 distinct products from which to choose. The assumption of perfect information may be very difficult to grant, but at least you can find more things to want!

Countless institutions, in pursuing their own self-interest, extol people to think like Rational Economic Man. Corporations that want to sell us something, the media that assist them in the act, and the advertising industry that packages the message all offer powerful examples of institutions trying to coerce people to think like Rational Economic Man.

Each of these institutions urges us to label ourselves as *consumers* first (as opposed to citizens or even just people) and reverently promotes the exaltation of that word. Government, with constant reference to consumer rights and consumer protection, is equally complicit.

In recent years, it has become fashionable for some political candidates to structure their messages to voters along increasingly narrow lines of self-interest. Voters should want to pay fewer taxes, receive more services, and vigorously seek to ensure that the "economically unworthy" not get to share in the "good life" or get something for nothing. And, of course, the politicians determine who is and is not "economically worthy," even while ascribing their choices to voter preferences.

The vision offered is almost never one that includes the long-term interests of posterity or the system, but rather is primarily of the *self* and *now*. The operational question is, "What can I do immediately for your tangible material well-being?" (Witness the calls for cuts in taxes when government cannot finance what it pays for now, such as corporate welfare.)

Of course, some people in a mass society are unpredictable with respect to the model called Rational Economic Man. For example, they think of others (act altruistically), are concerned about the long-term survival of the social/environmental system (thus ignoring the model's time preference), and/or allow emotions and spiritual concerns (irrational thinking) to dominate some of their behavior. At best, they tend to be ignored by dominant institutions; at worst, they are considered a threat to the existing order of things.

Rational Economic Man and Sustainability

People do occasionally act capriciously; for example, when we make different choices under similar conditions, we thus appear "irrational."

Clearly, we do not—*and never will*—have perfect information about all the choices and consequences we face in using resources. We face this imperfect knowledge not only in mundane decisions about our consumption but also, and more particularly, when uncertainty surrounds decisions with long-range consequences. Such uncertainty can occur for a variety of reasons, including: (1) lack of complete information on supplies or endowments; (2) uncertainty as to the unintentional effects of extraction, processing, and utilization; or (3) simply because the scale of time involved is so great that we cannot easily imagine all the possible ramifications of our decisions and subsequent actions.

To people, intangible considerations, such as love, security, and self-esteem, often take dominance over material acquisitions, despite the frantic and secular appearance of modern life. Further, securing all kinds of benefits, both tangible and intangible, for others, often in a manner that appears altruistic, is commonplace if one knows where and how to look. Thus, either self-interest does not hold or it has been redefined in some complicated and communal manner.

Finally, immediacy does not always dominate day-to-day decisions. Planning for a retirement nest egg or next year's tuition can, at least at times, uncover behavior in which a person will apparently give up more now to secure a certain (partially intangible) package of benefits later. Securing something of value for one's children or grandchildren is an act that qualifies as both long term and altruistic.

The fact that financially measurable decisions *usually* resemble the assumptions of Rational Economic Man (while problems in complying with the expectations of the theory are more likely with intangible assets, whether personal or collective) only proves the point. If the scope of the decision is narrow enough, then standard economics is a wonderful tool. It is the intangible or impecuniary assets with long-term implications and poorly defined rights of private property, such as most environmental factors, that cause Rational Economic Man to make unfortunate decisions or even to be impotent.

In summary, the assumptions of Rational Economic Man are probably valid in a fairly narrow sphere, where the ramifications are clear, the options known, and the span of time short. All the variables of economics and human behavior, which must be held constant to render the notion of Rational Economic Man valid, do not stay fixed in place for long. In deciding whether to stay home and watch television or go out to dinner and a movie, standard "rational" economic analysis would apply wonderfully. Deciding whether to cut or save a forest is another matter. In such a situation, rational, well-meaning people might not even be able to identify their own self-interest.

Discussion of the paradigm called Rational Economic Man has suggested assumptions about self-interest at many points. Even though self-interest is a subfeature of Rational Economic Man, it is critically important in the development of the theory of market systems and thus merits separate consideration.

Self-Interest

Self-Interest and Economic Theory

Like some of the other individual features of Rational Economic Man, the notion of self-interest is fundamental to the assumptions upon which the theory of market capitalism is based. Economic models normally require a process through which economic actors in the model (such as firms and consumers or workers) seek to attain the best situation they can, given scarce resources (time or money).

In order to behave *rationally,* however, which means doing the same thing every time and thus being predictable, a consistent set of goals must be assumed in the economic theory. The chosen goal is, of course, material satisfaction within a framework of concern based solely on oneself. Because the dollar value of material goods is measurable, the conditions that describe the accompanying human activity can be mathematically formulated. *Self-interest is thus assumed,* so that the behavior of an individual can be predicted according to a consistent set of principles.

Because self-interest is assumed, the elusive notion of mathematical functions that describe only one person's unique tastes for utility or some other preference can also be assumed. It is as though likes and dislikes, appearing magically and fully grown, are assumed to be unique to that person and beyond question, regardless of how aberrant or strange they may be.

Although this elusive formulation is impossible to observe, it can be assigned mathematical properties (if the assumed behavior is consistent enough), and much elegant and rigorous economic theory emerges. In short, the consumer must strive consistently for something, and the most tangible element possible is economic goods and services.

But if people do not behave with singularly acquisitive attitudes, the models simply will not work. Again, the assumptions keep the models workable, even though consumers may end up in an uncomfortable behavioral world not of their choosing. This is a clear example of the tail wagging the dog, whereby the consumer is strongly pushed through such means as constant advertising to act in the interests of the model builder and not in concert with his or her own likes or dislikes.

Self-Interest in Practice

If you are engaged in an economic transaction, say purchasing a fish for dinner, you expect the merchant to ask a certain monetary price for the fish, and the merchant expects you to demand a certain qualitative value in the fish. Why? Because in the world of commerce, people are expected to behave according to their own self-interests. More accurately, you expect others to behave according to *your own idea of their self-interests.*

Now we are getting somewhere. Your expectations are nothing more than a theory you learned sometime, somewhere, somehow, and a theory is adjustable, expandable, and relearnable. But that theoretical notion, probably unexamined for some time, determines the relationship between you and the merchant. "No problem," you say, "because providing my dinner is the main reason I need the merchant, and the assumptions (like driving on the right side of the road) make things work. (A head-on collision is, after all, only a quicker catastrophe than starving to death.)

The anointment of self-interest as a "noble" human motivator has clearly supported the presence of individualism throughout the European-influenced history of the United States. Individual initiative and achievement have been revered in our American heroes, ranging from the stories of Horatio Alger of the 19th century to the lavish attention heaped on successful businesspeople of today.

No society in the world approaches the United States in its tendency to canonize people for the mere fact that they have made a great deal of money. Think for a moment; not only must a single-minded workaholic capitalist constantly pursue his or her own self-interest but also he or she must accept a ready-made structure of value that defines his or her own measurement of success as monetary and material. Most people can think of someone who fits this description.

Self-Interest and Sustainability

What is wrong with all this? In a sense nothing, because it is impossible to imagine our system throwing out self-interest. As a practical matter, it is comfortable, reassuring, and desirable that the systemic order allows us to go about our daily activities with the knowledge that it is in the basic interest of others to be in their places doing their jobs. As someone once said, "I want the farmer's and grocer's dinners to rely on them getting me mine."

However, operating on the principle that self-interest is generally desirable for the economic system is quite a different matter than contending that each individual's own definition of self-interest is entirely appropriate. This is true even for oneself, let alone how the unintended effects of such a definition may affect others. It is thus clear that the individual interpretation and awareness of exactly what is *really* in one's own self-interest is the fundamental question.

The specific risk is that cultural conditioning will lead us to define self-interest too narrowly, causing us to think we must compete with the very people and institutions for whom cooperation and coordination are a much wiser order of the day. If social/environmental sustainability is to be a viable option, self-interest must increasingly be culturally (collectively) defined, even though economic theory still takes a totally individualistic approach to the treatment of consumer preferences. Although this may sound unrealistic, there is some hope.

It is completely logical to assume that our "theory" (despite whatever problems we may observe with it) also frames the social relationship with family, friends, neighbors, and teachers, to name a few. The upshot is that self-interest cannot be divorced from an individual's concept of community. Placing importance on the health and well-being of the community, including the health of a given individual's many personal relationships, is therefore a realistic proviso for anyone's definition of self-interest.

It may sound trite, but a sustainability-related goal of education must be to emphasize that the health of one's own community is in one's own self-interest because one is inextricably a part of one's community, and ideally wants to be. This would merely require emphasizing the communitarian thread of the American monomyth as opposed to the individualistic thread.

In an almost evolutionary sense, we are drifting into a world of the 21st century, where the state of Nature is such that traits like sharing, cooperation, and coordination promise to bubble up as those individual characteristics that are most likely to contribute to the survival of our species. The theory of evolution indicates that such characteristics as sharing, cooperation, and coordination have always been somewhere in the gene pool, but to develop the perspective fully, we must briefly review the nature of the rampantly individualistic world of 19th century North America.

Nineteenth century North America was a time dominated by abundant natural resources (both unexplored and unexploited by the European invaders) in which a lack of people was defined (again by the European invaders) as the chief detriment to material progress. In economic terms, labor was the scarce resource.

It was also a time of almost mythically documented westward expansion over the continent during which the cultural details of the

American character emerged, many of which we still lionize today. The traits of the times favored expansion, growth, and dominance over Nature and the indigenous peoples. Acting as a trustee of the natural resources to ensure their long-range availability for generations of the future was simply not an important trait in this world of temporary superabundance.

It has taken much of the 20th century to complete the expansion and evolve to a largely urban industrialized society and economic system. We have continually and aggressively developed and implemented more sophisticated "space age" technology and have rationalized this movement with a constantly adapted version of the cultural mythology left over from the 18th and 19th centuries. Thus, it is not surprising that the continual emphasis on growth has caused us to experience the problems that inevitably attend the movement toward, and in some cases beyond, the maximum rates of resource use that are consistent with the long-term carrying capacity of various resources and resource systems.

A comfortable transition to the 21st century would thus have self-centered traits (such as acquiring, competing, and growing) evolving into sharing, cooperation, coordination, and sustaining. To move toward this type of a future, people must be so educated that the concept of self-interest is expanded and refined in the dimensions of both time and space. Self-interest need not and should not be eliminated, however; it is a powerful tool that must be used to consciously focus community development on a sustainable track through time. But what about the "invisible hand"?

The Invisible Hand

The Invisible Hand and Economic Theory

The "invisible hand" is not only a well-known principle of economic theory but also possibly the greatest single "sound bite" legacy of Adam Smith. It is included here not because you, the reader, are likely to be unfamiliar with it, but because there are subtle behavioral implications, which are vital to understand as we dissect the structure of economic method in quest of social/environmental sustainability.

In short, the invisible hand thesis contends that everyone, acting narrowly and competitively in pursuit of their own self-interests, does what is best for society at large, as if guided by an invisible hand. After all, who does not want to believe that he or she is best serving all others by aggressively maximizing his or her own material well-being? This seductive premise is at once the genius and the potential downfall of the social efficiency of the free market system.

The formal logic is that attainment of full efficiency in a market economy demands that all resources be allocated to their maximum use. This requires that components of production (such as workers with their labor) be employed in their best possible use and that final goods and services be somehow allocated to those who obtain the most satisfaction from their consumption.

By actively seeking the best possible jobs and by acting as a clever, informed, and almost competitive consumer, the individual tends, in an economic sense, to promote this happy situation in his or her own micro sense. Expanding on this notion, the combination of all the components of production and all individual consumers acting in the same way magically achieves a social optimum (normally referred to as *efficient resource allocation*) for the overall macro economy. The sum total of this supposedly "automatic" process amounts to the most powerful apology imaginable for the social virtue of allowing self-interest to reign as the sole motivating force of human beings and thus for the structure of private market capitalism as an economic system.

Indeed, the entire philosophy underlying free markets is that each party, the producer and consumer, brings something of value to the transaction, and they exchange these valuable assets in such a manner that each is better off. Logic dictates that if there are only the two parties in a given transaction, both increase their satisfaction through the exchange, and no one else is affected, then the transaction must automatically increase total social satisfaction.

Furthermore, this ties the invisible hand inextricably with the institution of private property and the notion of economic freedom, not to mention self-interest itself. The skills employed in the work force and/or the other components of production (i.e., land, capital, and so on) are all owned by private citizens who are free to make them available to the highest bidder. Hence, the notions of economic freedom and private property "lubricate" the invisible hand. It is quite a happy and unfettered world, but, as we shall see in the next chapter, only if you own something of value to begin with.

The Invisible Hand in Practice

It seems clear that the notion of laissez-faire (the doctrine of freedom from governmental interference) accompanied the acceptance and application of the invisible hand thesis. The thinking in the private sector of the United States, as the world's preeminent experiment with the philosophy of laissez-faire, has always cycled between healthy skepticism and downright hostility insofar as governmental regulation (or virtually any direct involvement) in the private economy is concerned.

In practice, therefore, self-interest as the dominant feature of the invisible hand paradigm has been treated more as patriotism than as greed. American economic history, anointed heroes, and of course politics all reflect this presumption. The major architect of economic health is the capitalist, who, by taking risks, pursuing ideas and dreams, and amassing capital (normally in the corporate form), creates jobs, products, and opportunities for everyone. Investment tax credits and trickle-down economics cannot be far behind. Material acquisitiveness has become nobility.

Little more needs to be said about the invisible hand in practice once it is recognized that it contains the central motivating force of that most central individual in a market economy—the capitalist. Entire schools (e.g., the Austrian school of economic thought) have extolled the virtues of maintaining economic freedom in order to retain and create the incentives assumed necessary in unleashing human entrepreneurial and inventive potential. And always, the personal desire to accumulate, prosper, and "get ahead" is the supposed desirable trait of the individual—the "goose that lays the golden egg" for capitalism.

Although the possibilities for adding to this line of thinking are virtually unlimited, it will prove more instructive to incorporate additional comments within the discussion related to social/environmental sustainability.

The Invisible Hand and Sustainability

A major problem is seen by again reviewing the assumptions of Adam Smith. In contending that the greater social interest was served by the narrow maximizing behavior of private people, he assumed unambiguously that the actions of seller and buyer (supplier and demander) affected only each other and that no third party was in any way affected. It was a much simpler world in 1776, however, even though the Industrial Revolution, which so fascinated Smith, was just under way and would soon complicate matters.

If this identity between public and private interest is to be the case, then assessed property rights to be exchanged must in fact be perfect, and the buyer and seller *are* therefore the only parties involved. Formally, economists would say that there are no externalities (unintended effects) as a result of the production, consumption, or disposal of a particular product and that the transaction has indeed directly increased total social well-being. But then Adam Smith's buyer and seller may as well be Robinson Crusoe.

In assessing whether the invisible hand is an accurate and workable principle for our modern society, economic theory itself suggests examining the assumptions that must hold in order to contend that a free market system is efficient. Let's ask some of the appropriate questions:

- Are our economic transactions free of external effects on others, including pollution or any other type of externality?
- Are property rights to all resources clearly owned, defined, and used (including air, water, biological and genetic diversity, and environmental amenities)?
- Are the long-term effects of decisions concerning the allocation of resources injurious to others (or even known)?
- Are the resources used as a result of the transaction permanently available so that the options of future generations remain totally open?

For over two centuries, we have been busily populating, urbanizing, and in general complicating our world to the point that Adam Smith might not even recognize it. Clearly, a crowded urban world with increasingly scarce resources—including space itself—is the antithesis of Robinson Crusoe's island and therefore the type of environment least likely to have no unintended, unforeseen effects on others. As such, the modern environment is the least appropriate type of environment for the thesis of the invisible hand. Again, this is not ideology speaking, but simple logic.

Communities, whether large or small, are by definition concentrated nodes of economic activity. It is here that people live, work, and recreate most closely with one another. Therefore, it is in communities where externalities (external effects; unintended, unforeseen effects; neighborhood effects; call them what you like) are most prevalent. This is an obvious measurable, physical fact and has nothing to do with politics, liberalism, or conservatism.

It may, however, amount to an almost "scientifically ascertainable" reason why urban areas are more often centers of a liberal hands-on attitude toward government as opposed to rural areas, which often tend to be more politically conservative by today's definition. Residents of either area in effect observe different worlds—one in which others are automatically involved, the other in which individualism and the apparent lack of effect of one's actions on others are assumed to be the norm.

Ironically, we may have *grown* ourselves into a world in which the philosophy that gave us that growth ethic is no longer applicable. We have evolved into a world of interactive interdependencies, where unbridled self-interest does not have much hope of identifying a socially desirable course of action due increasingly to our technological processes, lifestyles, and the nature of our many institutions.

Adam Smith may be rolling in his grave in lament of the deterioration of his hallowed construct of the invisible hand. But if, at the time he wrote, he had even begun to understand the logical prognosis for industrialism, and there is good evidence he did, he shouldn't really be too surprised, particularly where productivity is concerned.

Productivity

Productivity and Economic Theory

Productivity is defined simply as the output per unit of time for a particular input, such as the output per worker per hour of time put in. Its common application is to human labor. Who could possibly find fault, in theory or practice, with something called productivity? After all, the very word implies the maximization of the payoff or reward for effort expended. You get something of value from what you invest. This is the essence of a successful economic system (especially one based on free enterprise)—value received for value expended.

Incidentally, we can extend the notion of value given for value received to the analysis of public services and taxation. It is the general philosophy on which public finance is based, which is both appealing and deceiving, namely that those who pay taxes should receive the benefits of public services.

The deceptive nature of productivity stems from a corollary often drawn: if one perceives that he or she receives no direct personal benefit from the actions of government, he or she has a right to pay little or no taxes. (These threads are important in understanding the current rhetoric concerning policy and in developing the concept of sustainability in economic analysis and will be discussed separately.)

Productivity is closely related to the invisible hand thesis. If, for example, an individual owns something of value, specifically his or her talents and skills, then the ability to exchange these talents and skills in the labor market (i.e., work for a living) is dependent on that individual's productivity. An increase in skills, say through additional education or training, leads to a gain in productivity, which results in the potential willingness of an employer to increase the compensation for these skills.

Hence, increased productivity leads to a higher salary, which leads to a greater ability to command consumer goods and services, and—in a totally materialistic frame of reference—a higher standard of living. Therefore, in a series of often complex relationships, productivity supports economic satisfaction, and increased productivity supports economic expansion and growth.

So far, this is standard American thinking. But the plot thickens when we remind ourselves that human labor is not the only valuable resource that an employer seeks to hire. There are other factors related to production, such as land, capital, and various physical raw materials.

In general, the discipline of economics is designed to focus on the productivity of all inputs (output per acre, productivity of capital, efficient use of land, and so on), not just on labor. In order to explore the implications of this, we must focus on the variable called capital, which will prove useful if we are to uncover the crux of the previously discussed growth/no-growth tug-of-war.

It is standard dogma that the productivity of labor and the productivity of capital operate inversely to each other, provided all other components of productivity remain relatively constant. Put differently, the components of production, such as capital (the machine) and labor (the person running the machine), should be paid according to their actual contribution to the final, measurable output of the product. Thus, a technological adjustment (such as a new machine) that increases the productivity of capital (the increased output of the new machine versus that of the old) increases the relative share of the profit that the machine should be able to command. This means, of course, a decrease in the relative share of the profit that goes to labor.

As an example, if a worker is given a better machine, that worker *appears* to be responsible for more production in a certain amount of time (per-hour productivity has risen), when, in reality, the machine does more of the work compared to the worker. There is, however, a vital but little-understood feature of conventional economic theory with respect to capital and labor that requires a brief explanation at this juncture.

The feature we are talking about is the interaction between capital (machines) and labor (workers) as valuable components of production, which are normally analyzed as a ratio or a mix wherein the absolute amount of either is not important. Thus, when the producer (employer) substitutes capital for labor or labor for capital, the substitution does not result in either more capital or more labor per se, but rather in an increase in the relative intensity with which machines (capital) are used or an increase in the relative intensity with which labor (workers) is used.

The law of diminishing returns for a factor of production holds that the productivity of the last productive component (capital or labor) added will normally drop the more there is of that component. In other words, if a business hires more workers, the incremental increase in productivity of the last worker hired drops when compared with the productivity of the other workers. Therefore, because productive components (capital and labor) are compared with each other (and given the reality of the law of diminishing returns), the important practical conclusion is that the relative productivities of capital and labor operate inversely to each other.

This inverse relationship means that as more of either capital or labor is employed, the incremental productivity of the more abundant component drops as the incremental productivity of the more scarce component increases. Thus, the incremental productivity of capital increases as more labor is hired, while the incremental productivity of labor simultaneously decreases, and vice versa.

These realities of conventional economic analysis are no doubt largely behind the reasons why capital (management) and labor tend to define their fundamental economic interests in opposition to each other in the day-to-day world of economic institutions. It is thus in the interpretation of productivity where the link between theory and practice becomes critical as we seek to understand the effect the notion of productivity has had within our economically oriented society and ultimately relate it to growth and sustainability.

Productivity in Practice

Clearly, the more productive worker has an easier time commanding an increase in payment. But an equally inviolable principle is that the entrepreneur in the "real world" of business seeks to hire relatively more of the cheaper (and thus more lucrative) component of production (machines) and relatively less of the expensive one (workers).

This situation leads the entrepreneur to use less expensive machines in place of more expensive people, setting up one of the greatest ironies inherent in our free-enterprise capitalistic system: as productivity of workers increases over time and the economy is deemed successful in supporting people, strong incentives are set up for a counterproductive theme in the economic drama, whereby the businessperson or entrepreneur replaces productive people with machines. Therefore, *successful performance by workers tends to make employers want to replace them,* at least if workers want to receive the higher salaries that should, according to theory, accompany gains in productivity.

Enter growth and technological change, growth in particular. Without growth, capitalism would tend to careen wildly between boom and bust as the relatively cheaper resources are perpetually substituted for the relatively more expensive ones. Unemployment would tend to occur suddenly, geared to the speed of technological change, and the cycles in business would fluctuate unpredictably and severely.

Perpetual economic expansion would save us from all this, but it carries two prerequisites. First, there is need for a constantly increasing investment of capital to support the needs of continual economic expansion. Second, technological innovation must support constant gains in productivity and provide incentives to change the status quo mixture of capital and labor. This change is deemed necessary because economic theory makes the following assumption about efficiency: because inputs (in particular workers) are already allocated efficiently in a static state of technology, further gains in productivity must await advances in technology. The important point is the *mix* of capital and labor, not one or the other in isolation.

Thus, the story would end in a dynamic equilibrium—at least as long as the availability of resources and the environment would allow. Put differently, if resources are infinitely available, this set of theories would remain viable. The system we have constructed, emphasizing as it does a constant muddling toward higher rates of growth, would be the permanent state of affairs, even though minor differences in ideology would come and go.

Productivity and Sustainability

We are confronted by three questions in our attempt to integrate the concept of social/environmental sustainability with current economic theory: (1) Is our economic construct that emphasizes continual growth possible? (2) Is our economic construct that emphasizes continual growth desirable? (3) Will the relentless acquisitive nature of our economic construct work harmoniously with respect to social/environmental sustainability?

As already stated, we are convinced by other data that our economic construct, which emphasizes continual growth, is not sustainable, although we choose not to concentrate on the proof of that here. The second and third questions are both intriguing and bound intrinsically to the notion of social/environmental sustainability.

Adapting the notion of productivity in the service of social/environmental sustainability requires, first and foremost, that productivity can and must apply to all economic inputs. As we have implied, the notion of productivity has been overused as it is applied to labor (human productivity) and underused with respect to other components of production, such as capital (energy and the land's productive capacity). It is these "other" that now threaten to become scarce.

Scarcity can occur in two ways: running out of something (e.g., oil) and overusing something (e.g., polluting the land or air). Either way, one could say that the rate of use of a given resource, be it physical or intangible, in today's world threatens to exceed the carrying capacity of that resource if the current level of consumption continues unabated.

In a world threatened with scarcity, the general principle of first concern must be the *sustainable* productivity of threatened resources. We disagree with the notion that economics automatically deals intrinsically with the sustainability of scarce resources through the pricing system and the resultant reallocation of those resources. On the surface, however, this disagreement would appear to be poorly based.

The energy crisis itself appeared to demonstrate the principle of first concern. Prior to the Oil Shock, there was little concern for efficiency in the use of available energy, since oil was historically cheap. Economic theory is clear under such conditions: If oil is cheap, use more of it relative to other kinds of energy. This is what the American economy did—with total abandon—and called it an increase in labor productivity.

If both technological innovation and public policy concentrate on enhancing the productivity of a threatened resource, other, more abundant resources are relegated to lower levels of awareness or concern. After all, "productivity" appears to be going up for the threatened resource as we lessen the demand for it by using other, more abundant resources in its place. The benefit should therefore be higher levels of employment in conjunction with the more abundant resources.

For instance, shortages in energy will result in slightly lower wages, but maintaining levels of employment should be easy as long as energy and labor are interchangeable components of production. If, however, we remain hooked on "labor-saving" devices, where gains in productivity at the expense of labor are necessary to support this habit, scarcity in other physical resources will leave us perpetually vulnerable to unemployment and other economic crises. The upshot is recurrent instability and continually worsening distribution of wealth and income, which brings us to technology.

Technology

Technology and Economic Theory

Few topics will stimulate more spirited debate among self-appointed social critics than the role of technology in modern society. To an economist, technology is the state of the art in production. It defines the ways in which people and machines combine to create products. Thus, since productivity is output per person-hour (or, in general, per unit of time), very little can be said about the innate productiveness of a human being until the level of technology is specified.

In effect, human labor is just one of a variety of resources marshaled and combined in some format to accomplish the act of production. The result, of course, is that the formal discipline of economics views the inputs (e.g., capital and human labor) primarily in combinations and ratios of one another and thus tends to focus on the product itself as the object of its analysis.

Although it perhaps goes too far to conclude that this focus actually demeans the role of human beings in the day-to-day acts that serve to meet our material necessities, it does create a formal indifference to the distinction between human and machine. From our vantage, the entrepreneurial decision will probably be to continue substituting the relatively less expensive labor-saving productivity of machines for the relatively more expensive labor-intensive productivity of human beings. As such, productive processes will become unerringly more capital intensive and less labor intensive.

Technology in Practice

The role of technology has come to be viewed as the avenue to all "improvements" and "progress" in economic affairs. Since the combination of human and machine creates products as output, a steady state in the technological arena is tantamount to a stagnant economy. In other words, without new technology, we cannot have gains in productivity, and without increases in productivity, the seductive door to ever-higher standards of living remains shut.

The combination of technology and free markets has played such a specific role in the economic history of the United States that a distinct ethic, known as technological optimism, has emerged over the years. This ethic has become so widely assumed and well developed that it has virtually become a part of standard economic theory, not to mention our common mythology. In fact, technological optimism is totally central to the economy versus the environment dilemma and the growth/no-growth tug-of-war.

In its simplest form, technological optimism merely contends that technical know-how will bail modern industrial society out of any problem that may come along. In a nutshell, necessity is the mother of invention. This is a straightforward and familiar concept and represents a position we have all taken on many issues at one time or another. Its devotees are legion and its credentials impeccable. Given our purpose here, the cluster of ideas represented in the notion of technological optimism merits our attention, and the following scenario may well sound familiar.

Assume that some sort of constraint or tension arises in a society. It may be simple, such as a desire for a new or better product, or it may be more serious, such as a major shortage or wartime threat. Assume that a product is in short supply and is traded in some market. Now, due to the short supply (and presumably high demand), the price goes up. The increase in price acts as a signal that someone finds this product valuable and is willing to reward anyone who is able to satisfy the demand.

In a simple incentive system (remember, self-interest is alive and well here!), someone acting in his or her own self-interest develops a new alternative or finds a new resource and is richly rewarded for his or her ingenuity in bailing the system out of its bind. Technological innovation is therefore at the heart of the ability to forestall the crisis. The details may vary, but the general litany is familiar: "Technology will bail us out."

Labor unions, for instance, always focus on measurable gains in productivity as they bargain for higher wages. The institutional fact is that both sides of the bargaining table are essentially assuming, although it may not be explicit, that the quality of the machines is the important factor. This assumption may go a long way in explaining the declining effectiveness of the formal labor movement in our current late industrial era.

Thus, despite much rhetoric about quality of the labor force and the value of human beings, there is a deep-seated feeling that the quality of the *capital stock* (i.e., the machines) is the *real* determining factor in achieving prosperity. Finally, changes in the quality of this capital stock, through technological innovation, are seen as crucial in creating the dynamics we call economic growth.

Technology and Sustainability

The problem with technology is that it has not always been consciously directed toward serving the necessities of human beings, a thought captured in the reflection of author Havelock Ellis: "The greatest task before civilization at present is to make machines what they ought to be, the slaves instead of the masters of men." Technology has instead developed willy-nilly within the realm of scientific and engineering possibilities. Some technological developments have therefore admirably served the necessities of human beings; some have not.

In the early stages of the Industrial Revolution in Great Britain, the Luddites stormed the factories and smashed the machines, which they blamed for supplanting their jobs. A "neo-Luddite" attitude will not guide us successfully into a sustainable future and certainly is not even remotely acceptable to the powers that be.

The goals and objectives to be met through and with technology must be consciously chosen from a sound basis of core human values and in light of a detached, dispassionate assessment of the state of the world and its inventory of available resources and their ecological sustainability. Once again, conscious judgment concerning durable human values cannot and must not be avoided. But remember, in Nature's scheme of things, the principle of cause and effect is impartial, no matter how much we might wish it otherwise. Having thus completed our discussion of economic terminology, what principles, if any, availed themselves as we considered social/environmental sustainability?

General Principles

Now that each of the terms has been covered in the same format, one all-important question remains to be asked before proceeding to the next chapter. In examining the third section of each of the five concepts, what general principles arose as we discussed adapting the terms to the concept of social/environmental sustainability? We suggest the following:

- Rampant individualism must be supplanted in both economic theory and practice by other-centered sharing, cooperation, and coordination.
- Human values must be prominently included, not excluded, in both economic theory and practice.
- A long time frame must be consciously adopted, which means that the well-being of future generations must be purposefully included in economic theory, planning, and practice.
- Communal (group) visions, goals, and relationships must be addressed, accepted, and nurtured in both economic theory and practice.
- Self-interest must be continually subjected to reassessment in its broadest sense in both economic theory and practice and assessed in terms of its effects on the generations of the future.
- Human values (such as the quality of human relationships in both the workplace and the community at large) and nonmaterial assets (such as the beauty of a community's surrounding landscape) must be given a more prominent role in economic theory and practice.

These principles are the initial foundation for a theory of social/environmental sustainability and thus provide the basis for proceeding to the next chapter. In addition, while it might initially seem that many terms need to be addressed in order to be comprehensive, we hope it is clear that this is not the case.

The terms we have chosen are certainly fundamental. Although others could have been used, they would have inadvertently invoked the ones we addressed, even as it is apparent that the ones we addressed raise still other potential terms. Economic methodology is systemic and thus interactive and interdependent; therefore, we need to examine the terminology only until we are convinced that the general principles are uncovered. We hope our approach has made the complexities a bit simpler as we put ecology and economics together, where they rightfully belong. And it is now time to deal with the concept of distribution in economic parlance.

DISTRIBUTION IS THE KEY TO ECONOMIC SUSTAINABILITY

7

As we begin this chapter, it is important to realize that both ecology and economy have the same Greek root, *oikos,* a house. Ecology is the knowledge or understanding of the house, whereas economy is the management of the house, and *it is the same house.* And a house divided against itself cannot long stand. Yet, it has been the assumption of our society that if we manage the parts correctly, the whole will come out right. Although evidence to the contrary now comes from all directions, our systems of knowledge and management are still structured around this assumption.

"Call a thing immoral or ugly, soul-destroying or a degradation of man, a peril to the peace of the world or to the well-being of future generations," wrote E.F. Schumacher, "as long as you have not shown it to be 'uneconomic' you have not really questioned its right to exist, grow, and prosper." This quote points out the untenable divisiveness that weakens the house inhabited by ecology and economy. The question, therefore, is how to heal the current division in the house. Addressing economic theory is a major step.

AN ECONOMY IN A NUTSHELL

The role of an economy is to produce "things" to meet the material needs (in standard economic theory, read *wants*) of people. It is standard introductory dogma that any economy must answer three basic questions: What? How? For whom? In other words, an economy must decide: (1) which goods and services are to be produced, (2) with which combination of technologies and resources, and (3) who gets to use and enjoy the results in what proportion. To introduce some slightly more formal terms, the economic process answers questions of production and distribution of products and services.

It must be noted here that a free enterprise capitalistic economy attempts to answer all these questions through the mechanism of the market, supposedly in an impersonal and value-free manner with the aid of a "benign competitiveness." We will shortly return to a more complete discussion of the market system in this context, but understanding the complete picture requires some additional background information.

Early economic thinkers made considerable use of the word "subsistence," which we can define economically as meeting bare human necessities. It was as though the grand purpose of an economy was to free humankind from the Dark Ages or even the cave people from the status of hunters–gatherers. Gradually, however, as the Industrial Revolution unfolded and technological change became standard fare, the success of an economic system came to be measured by the degree to which it could elevate its participants above the perceived subsistence level.

The "materialistic hierarchy" goes something like this: Primitive is bad. Developed (= industrialized) is better. Rich is best. Can an almost religious fervor for economic growth be far behind?

Today, of course, almost any economy that is even semideveloped is able to produce enough to "sustain" its citizens at something measurably above the perceived level of pure subsistence. Supposedly, therefore, the question of current *production* is not the major economic challenge, but what happens when there is a surplus?

The Disposition of Surplus

An intriguing way to categorize and assess any economic system is the manner in which it handles "surplus." Surplus can be defined as production (and, of course, subsequent use or consumption) over and above the level of perceived subsistence. Moving beyond subsistence therefore involves the creation of surplus. Hence, when an economy gains the ability to generate products and incomes in excess of the necessities perceived as subsistence, that excess is distributed.

Pursuing this point briefly can demystify the formal disciplinary attitude toward the field of economic development. Producing for subsistence at its most elementary level involves simply producing enough to satisfy the necessities of basic food and shelter, which explains why an agrarian economy is often seen as "underdeveloped." Industrial production is therefore above the level of subsistence by definition, because it is seen as producing surplus. Thus, in the broad scheme of "richer is better," industrial production is thought to be at a higher level of "development" than is agricultural production.

More than any other factor, it is the method of distribution that characterizes economic systems. Some broad examples will underscore this vitally important point. Early civilizations, such as the Egyptians and Romans, employed slaves to create surplus, maintained the slaves at the level of subsistence, and kept the surplus they produced to support the upper classes of Egyptian and Roman citizens in lavish lifestyles.

In Europe, on the other hand, little surplus existed during the Middle Ages. Due to Europe's complicated feudal system and the marked influence of religious thought, what surplus did exist found its way to the Church and/or the ruling monarch and nobility, who were seen as earthly "representatives" of the Kingdom of God.

Then, in the mid-18th century, the British Industrial Revolution substituted machines, which employed the power of fossil fuel and steam, for slaves, who used essentially the power of wood as well as solar power. This all-important transformation of an economy from a primary source of energy, which is a renewable *flow,* to one that is dependent on a secondary source of energy, which is a depletable resource or *stock,* was accompanied by the ascendancy of a new and major actor in the economic drama, namely the *capitalist.*

The existence of *capital,* as the machines came to be called, naturally gave birth to the capitalist or person owning the machines and, of course, to *capitalism,* which simply put the appropriate label on this system of private ownership of the components of production. Due to the harnessing of these seemingly wondrous new technologies, it was now possible to create surplus at an almost alarming pace.

As a result, there arose questions about the distribution of the surplus, which would not have surfaced if virtually everyone had been at the level of subsistence. But the capitalist was clearly above the level of subsistence. It was therefore concluded that the capitalist should get to keep and use most of the surplus because he (and it was virtually always a "he") owned the machines that were apparently responsible for the dramatic increases in both production and the ever-widening gap between the requirements of subsistence and actual excess.

The story does not end here, however. Tracing the parallel development of the dominant economic systems and the economic thought describing those systems leaves us with the capitalist keeping most of the surplus in the early 1800s—and squarely in the middle of the classical economics of Adam Smith, David Ricardo, and Thomas Malthus.

But the pollution and squalor of the working conditions in the early European factories were apparent to Karl Marx, the most heretical member of the classical economic school. Although Marx, as did his compatriots, accepted the labor theory of value, which contends that human labor is the basic underlying source of all productive capability, he nevertheless saw the workers being exploited.

They were customarily paid only enough wages to meet the required standard of subsistence, despite the fact that they (certainly in conjunction with the emergent machines) were fundamentally responsible for the dramatic increases in the output of commodities. To Marx, therefore, the capitalists were unfairly expropriating the surplus, and the resultant pattern of distribution was unfair.

Distribution

By now it must be clear that, by any stretch of the imagination, distribution is the most problematic of the basic elements of economic theory, especially as one considers social/environmental sustainability. Who gets what? How much do they get?

On a planet offering a shrinking array of resources, the specter of scarcity is ever-increasing. The time-honored "expand-the-pie" dictum (with its trickle-down mechanisms) thus promises to be less viable as a method of offering economic hope to those near the bottom of the heap, whether at the local, state, national, or international level.

Arguments in defense of continual economic growth as the ultimate answer to questions of distribution will contend that inequality is a necessary effect (the other side of the coin, if you will) of the accumulations of capital required for the large investment projects deemed necessary to keep the economy moving. In their most virulent form, these arguments contend that concentrations of wealth (read that "inequality") demonstrate to the worker, the underclass, or those just starting out that they, too, can reap the benefits of the "land of opportunity" if only they are diligent and hard-working. Recitation of this "fairy tale" could continue ad infinitum.

Although the fairy tale may have been more or less true for some people throughout the economic history of the United States, it has not been true for many others, no matter how hard they have worked. In addition, increasing scarcity changes the accepted dynamics, which is probably a major reason Western industrialism is so slow in accepting and reacting constructively to the onset of the Era of Limits. In short, inequality becomes less defensible in the face of ever-increasing scarcity as merely a "semi-undesirable" effect of the accumulations of wealth necessary to fuel the engine of the capitalist market economy.

It is helpful to review briefly the conventional wisdom of modern market economics on the issue of distribution. The economic system employs all sorts of resources or components of production. At this point in the discussion, however, we are focused on the human component of production, or labor, since the primary issue with respect to the distribution of income is the incomes of people who work within the system. The question thus becomes how the products and services generated by an economy (usually symbolized by the gross national product) are supposedly reallocated or distributed to the participants in that economy, most of whom contributed in some way to its creation and maintenance.

Free enterprise market economic theory is crystal clear on this point: The valuable products of an economic system are reallocated to the components of production (be they people or machines) in the proportion to which either the people or the machines contributed to the making of the products. In other words, both people and machines are paid in accordance with the incremental increase in production for which each is considered to be responsible. Whoever contributes to the production of the gross national product gets to consume it in like measure.

Despite the superficially workable nature of this "benefits received" philosophy, it is fraught with perils when examined from an overall philosophical perspective. Four points will serve to lay bare for examination the structural underpinnings of capitalism:

1. **Decisions are made impersonally by the "market."** Supposedly, one of the beauties of a market system is that the reallocation of products and services to people in the system occurs impartially in a "value-free" manner. Here's how it works: An individual secures employment in some productive enterprise and is paid for what he or she is worth, given the nature of the job and the skill and training required. Then, the individual is free to

spend that income in the marketplace, where the fruits of his or her labor are offered for sale to this same individual (now acting as a consumer) who produced the product or products in the first place. The depiction of this process in textbooks on economics is often known as the "circular flow economy."

The touted advantages of the circular flow economy are that complete freedom exists both to choose a job and to spend the income generated by that job. No one tells you where to work or what to consume. You "vote" with your dollars according to your own preferences. In turn, decisions are made about what to produce in response to how you vote with your dollars or purchasing power, as it were. The market system can thus be said to be "value free," other than the exercise of the personal values driving individual choice.

2. **If you do not produce something, you do not get to consume.** An individual worker/consumer is paid if he or she creates something of value for the market and is not paid if he or she does not. Given the market's way of allocating items for consumption (purchase) by those who produce them in the first place, this means that an unemployed person does not get to consume.

   Supposedly, some people in a mass economy fall through the cracks, and either cannot work (are disabled or handicapped) or cannot find a job (are unemployed). While it is hoped that unemployment is temporary, it is nevertheless assumed that the unemployed have little training and few skills and would thus have low increments of productivity. Although capitalistic thinking reluctantly acknowledges that allowances must be made for these "problems" (= disabled, handicapped, and unemployed people), much controversy exists over questions of the nature of support they deserve and how much.
3. **Some inequality exists because of innately unequal endowments of talent, skills, and training among people as workers.** Since all people are not endowed with equal skills and ability, some human workers (= some kinds of labor) are more valuable to the economic process than others. Hence, wage and salary differentials exist.
4. **Competitive conditions exist at all levels of the economic process.** Much has been said and written about competition. In this context, it is quite simple. The existence of competitive conditions ensures that several things occur: (1) the best worker for

a particular job finds his or her way to that job; (2) the worker is employed in the best possible manner, given his or her talents, skills, and training; (3) this will occur at the lowest possible cost to the employer and the highest possible salary for the worker; and (4) as goods and services are distributed, the consumer has the largest possible choice in "voting" with his or her consumption expenditures.

All of this amounts to competitive conditions in the markets, where people are hired and final goods and services are exchanged as the final products of the market. Workers/consumers compete for the best jobs and businesses/suppliers compete to sell their products and services to consumers. Overall, it is a process much revered in the literature of free enterprise and market economics.

The sum total of all these conditions amounts to an omnibus state of affairs to which economists often refer: efficiency in overall allocation of resources. Efficiency in overall allocation of resources supposedly ensures that the resources available to a particular economy are employed in creating the maximum possible amount of human welfare.

This "state of efficiency" is attainable for any economic system, be it large or small, rich or poor, complex or simple. It refers to an economy's use of its resources (employment, harvesting, mining, depletion) for the production of goods and services for distribution and ultimate consumption to satisfy as efficiently as possible human desires through consumption. But, of course, this says nothing about equity or fairness of the resultant patterns of distribution and wealth. The fairness of the distribution and wealth is supposedly a political question, not an economic one.

A sustainable economic system demands that decisions about distribution be made *consciously* by human choice rather than unconsciously by the market. Sharing in some manner must replace "expand the pie and we can ignore sharing," which brings us to the notion of economic feasibility.

ECONOMIC FEASIBILITY

In order to bring this discussion of values and economics concretely into the present and tie it into the broad historical framework, let us go to a seemingly strange place—the concept of economic feasibility. The reason this may seem strange is that the notion (and test) of economic feasibility appears on the surface to be technical and value free, which is precisely the point of this and other examples we could have chosen.

Economic feasibility *appears* to be value free. To clarify this point, we must consider not the concept itself but the modern framework within which it is employed.

The test of economic feasibility is often elevated to the status of a "public religion" in the world of practical affairs. If some proposed action or project is shown to be, or even accused of being, "not economically feasible," then it must certainly be facing a perilously short life span in the material world.

What can be deduced about the mindset and core values that lead people to place such prime importance on the notion of economic feasibility? As an analytical tool, economic feasibility provides an easily accessible window into the room that houses the collection of personal and societal values involved in defining the difference between continual growth and sustainability. The analytical path can take two possible directions: (1) empirical and technical and (2) value based.

Empirical and Technical

In order to determine feasibility, are we sure the analysis is done correctly and in a comprehensive and unbiased manner? Have we considered all the *real* variables? Have we taken into account the interests of everyone who is affected? And finally, once the relevant variables and methodology are chosen, have we accurately measured all these effects? The questions here are very similar to a benefit/cost analysis, wherein the analyst must make sure the technical research is done both correctly and comprehensively.

These questions, while important in their own right, do not ask the "why" questions (the questions dealing with underlying values) and thus cannot be the main concern at this point. After all, the ethic with which economic growth takes place ultimately boils down to human choices—not technological or scientific imperatives.

Value Based

What dominant set of social values leads us to think one way or another? Are they legitimate, and do they accurately represent the people who are the building blocks of the economic system? What in our human makeup leads us to conclude that "economically feasible" projects will advance the well-being of our communities or society more rapidly than proposals that do not meet those particular guidelines?

We must be as objective as possible, if a collective need to emphasize such things as economic feasibility arises from values that are inextricably and unquestionably rooted in so-called "human nature." Should that be the case, there is probably little we can or need do. Of course, values do not exist in isolation, but rather as clusters of compatible attitudes and deeply held beliefs. We contend that strong faith in the virtues of economic feasibility is but a preliminary step on the journey of allegiance to a fully developed ethic of economic growth.

This question of the universality of human values underlying economic behavior is neither new nor easily addressed directly. For us, the appropriate detour is to revisit the paradigm of Rational Economic Man. As stated earlier, this construct summarizes the economic assumptions held to be inherent in human behavior.

To the mainstream economic practitioner, Rational Economic Man summarizes what people are and how they behave. Let us therefore turn to some observations about Modern Economic Society. Rational Economic Man (or REM) exists in Modern Economic Society (or MES). In short, REM has found himself in a MES!

Economic Culture: Connecting Theory and Practice

The discipline of economics and its determined followers have certainly been central in creating the ethic of economic growth and its many ramifications. This "economic-health-comes-first" philosophy, whatever significant advantages it may have bestowed on human societies, clearly has been responsible for promoting the acquisitive, "me-first," "get-ahead-at-any-cost" lifestyle often noted by contemporary social critics.

As the title of this chapter suggests, the goal is to examine what it would take to move economics a bit toward sustainability. This is no mean feat, since, as the accepted underlying premises indicate, the discipline of economics will only move in this direction kicking and screaming! There are, however, some additional comments necessary in order to imbed this task in the groundwork of Chapters 5 and 6.

In addressing important questions surrounding the economics of sustainable communities, most people experience a "problem of connectedness" as they seek to develop their own rules of operation with respect to their use of the planet's resources. The problem stems from the fact that most of the critiques on the "limits to growth," although excellent, are on a global scale.

If, therefore, an individual seriously assesses much of what has been said and written, he or she must still make the connection between these macro critiques of growth-oriented people throughout the economy and their own individual behavior and personally held beliefs. This brings us back to the WNRCP hierarchy of Chapter 5, which emphasizes that the flow of resources occurs within an interrelated nested hierarchy that can be depicted as follows:

World
National
Regional
Community
Personal

To restate the problem of connectedness in these terms: How does an individual assess the data from the perspectives of the world and nation and act on them in the spheres of the community and personal?

"But what," you might ask, "about the regional level?" Given the typical situation in the United States, this would amount to the state, and the question can be rephrased: What guidance can a state, such as Oregon, offer to committed communities and individuals in promoting the goals of social/environmental sustainability? Our answer is that the state can do quite a bit.

The state's role is especially valuable (as the WNRCP hierarchy suggests) in serving as a connector, a focus, and a translator between the national and community (policy) levels on the one hand and the global and personal (philosophical) levels on the other. Ideally, the state is simultaneously close to the values of its citizens but also capable of maintaining a broad societal view.

Be that as it may, it is our strongly held belief that we *must not* look first to the state level. Communities can teach the state more in the early stages of addressing the issue of social/environmental sustainability than the other way around. Ultimately, it must be a dynamic partnership.

To return to the main thrust of our argument, the challenge of social/environmental sustainability is to derive an appropriate personal philosophy with which to focus personal actions and community development as one considers much that has been written and said over the last 25 years. The magic number can be put at about 25 years, since this dates back to such events as the first Earth Day in 1970, publication of the original *Limits to Growth* study in 1972, and the Oil Shock in 1973, which ushered in the energy crisis.

In short, the early 1970s began the era of generalized widespread public concern with the ultimate reality of what Kenneth Boulding has called "finitude." This period saw expanding acceptance of the belief (fear, some would say) that there exists a global carrying capacity for human beings because of the way in which the human population affects the resource systems of the world. It also brought out the necessity of addressing the "problem of connectedness" because for most people it is a lack of connectedness that makes it difficult to find answers when a committed individual says, "Okay, I'm convinced. What can I do here and now?"

A traditional economic theorist might tend to state this challenge as follows: "Existing critiques of growth-based economics have largely been at the macro or systemic level. They need to be scaled down to the micro or individual level." Although this is a beginning, it is inadequate. Further, it is dangerous because it contains the self-protective language that retains the emphasis on supposedly "value-free" analysis, and thus protects the intellectual and practical investment in traditional economic theory.

Microeconomic theory is fine, but still purports to be a positivistic and value-free analysis in which the emotional sterility of an assumed acquisitive, self-interested value base disregards the complex range of personal motivations implied in thoughtfully advocating and approaching the criteria of sustainability.

The analytical base must begin at the micro level, which means that implications must be developed for individual firms and consumers as opposed to the level of the world and national economies. In addition, the analytical base needs to be scaled down to the *personal* level and involve the *values* of the individual. The often-heard dictum "think globally, act locally" is an example of the attempt to address the "connectedness" between the world level and that of the personal in the WNRCP hierarchy.

People are confronted with good evidence that the planet faces serious problems of scarcity some time in the near future—unless *we change our thinking and our behavior.* Believing this, people face the realistic fact that options for actions overwhelmingly present themselves in their own communities and among people and local institutions with whom they interact daily and are quite familiar.

Indeed, for most of us, the actions we take within our own communities and among the people we know are the only way in which we will ever be able to promote the principles of a sustainable community. Although the links between the global and personal levels do exist, they are not obvious, so the tendency is to continue life and business as usual.[20]

As an example, a man living in Seattle, Washington, is struck by the loss of biodiversity as a result of deforestation in Brazil and wants to know what can he do. A woman in Tillamook, Oregon, wonders what she can do about the pollution of the Pacific Ocean. Children in a school in New York City discuss depletion of the ozone layer and wonder what it means to their future.

Somehow, growing an organic garden, using phosphate-free detergent, and applying sunblock during outdoor recreation, while "connectable," hardly feel like adequate responses to the apparent magnitude of the problems. What other options might there be, short of completely committing one's life to some cause? Exploring this question will take some time, and assisting people to begin this task in their own local communities is an overriding purpose of both this book and a previous one on sustainable community development.[20]

A final problem with addressing (and, if necessary, suggesting adjustment to) the role of economic philosophy and theory might already be apparent to the discerning reader, namely the connection between environment and economics. The whimsical "examples" given above began with global *environmental* problems, not *economic* ones. Nonetheless, be assured that the loss of biodiversity, pollution of the Pacific Ocean, and global warming are all directly caused by human economic activity—activity that, we will be reminded in no uncertain terms, is designed to feed and house the peoples of the world and in a number of ways to make life better for us all.

Here we confront a major irony in the consideration of our growth-oriented economy and public policies: The activities promoted as improving the well-being of people end up being directly and indirectly responsible for the degradation of natural and human-made systems. These activities thus cause considerable loss of human satisfaction, both current and future. So we end up degrading human well-being and maybe even seriously threatening the global carrying capacity of the human population as we energetically and unconsciously *promote* human well-being in economic isolation of the ecological systems on which that well-being depends.

We now finish this chapter in the same vein in which it was begun: The really important questions about the theory and practice of economics are those concerning human values and distribution of goods and services.

"Who gets what?" is the only important question, since "How much can we produce?" (a question essentially for engineering and technology) will inevitably lead to an answer of "not enough." The answer "not enough" assumes, of course, that the realities of finite resources are not reconciled with the impossibility of the "more-is-better" mindset about the distribution of goods and services encouraged by the Western industrial ethic of continual economic growth.

"Who gets what?" is important not only in terms of economic theory and practice but also in terms of human relationships. What economic theory and practice ignore is the *quality* of interpersonal human relationships. This is a tragic oversight because it is the quality of interactions with other humans and with our environment that not only gives life its value but also makes sustainable community development a real possibility.

8 HUMAN RELATIONSHIPS ARE THE SOCIAL GLUE OF A COMMUNITY

Relationships are the strands in the web of life, and there is no escaping the web. Relationships are all there is. In fact, life is relationships, and relationships are life. One cannot exist without the other because Nature's design is a continual flow of cause-and-effect relationships that precisely fit into one another at differing scales of space and time and are constantly changing within and among those scales. Every day of our lives is therefore about learning how to relate to ourselves, one another, and our environment.

When dealing with scale in relationships, scientists and economists have traditionally analyzed large, interactive systems in the same way that they have studied small, orderly systems, mainly because their methods of study have proven so successful. The prevailing wisdom has been that the behavior of a large, complicated system could be predicted by studying its elements separately and by analyzing its microscopic mechanisms individually—the traditional, linear thinking of Western society that views the world and all it contains through a lens of intellectual isolation. During the last few decades, however, it has become increasingly clear that many complicated systems, which are seemingly chaotic, do not yield to such traditional analysis.

Instead, large, complicated, interactive systems seem to evolve naturally to a critical state in which even a minor event starts a chain reaction that can affect any number of elements in the system and can lead to a catastrophe. Although such systems produce more minor events than catastrophic ones, chain reactions of all sizes are an integral part of the dynamics of a system. According to the theory of "self-organized criticality,"[24] the mechanism that leads to minor events is the same mechanism that leads to major ones. Further, such systems never reach a state of equilibrium, but rather evolve from one semistable state to another.

Not understanding this, however, analysts have typically blamed some rare set of circumstances, some exception to the rule, or some powerful combination of mechanisms when catastrophe strikes, again often viewed as an exception to the rule. Thus, when a tremendous earthquake shook San Francisco, geologists traced the cataclysm to an immense instability along the San Andreas fault. When the fossil record revealed the demise of the dinosaurs, paleontologists attributed their extinction to the impact of a meteorite or the eruption of a volcano.

Although these theories may well be correct, systems as large, complicated, and dynamic as the Earth's crust, an ecosystem, a civilization, or a society can break down under the force of a mighty blow as well as at the loss of a tiny, hidden safety wire, such as that mentioned earlier in the engine of a helicopter. Large, interactive systems perpetually organize themselves to a critical state in which a minor event can start a chain reaction that leads to a catastrophe, after which the system will begin organizing toward the next critical state.

Another way of viewing relationships within systems is to ask a question: If change is a universal constant in which nothing is static, what is a natural state? In answering this question, it becomes apparent that the balance of Nature in the classical sense (when disturbed, Nature will return to its former state after the disturbance is removed) does not hold. For example, although the pattern of vegetation on the Earth's surface is usually perceived as stable, particularly over the short interval of a lifetime, the landscape and its vegetation in reality exist in a perpetual state of dynamic balance (disequilibrium) with the forces that sculpted them. When these forces create novel events that are sufficiently rapid and large in scale, we perceive them as disturbances.

Perhaps the most outstanding evidence that relationships in a natural system are subject to constant change and disruption rather than in static balance comes from studies of naturally occurring external factors that dislocate ecosystems. For a long time, says Dr. J.L. Meyer of the University of Georgia, we failed to consider influences outside ecosystems. Our emphasis, she said, was "on processes going on within the ecosystem" even though "what's happening [inside] is driven by what's happened outside." Ecologists, she points out, "had blinders on in thinking about external, controlling factors," such as the short- and long-term ecological factors that limit cycles.[25]

Climate appears to be foremost among these factors. By studying the record laid down in the sediments of oceans and lakes, scientists know that climate, in the words of Dr. Margaret B. Davis of the University of Minnesota, has been "wildly fluctuating" over the last two million years, and the shape of ecosystems with it. The fluctuations take place not only from eon to eon but also from year to year and at every scale in between. "So you can't visualize a time in equilibrium," asserts Davis. In fact, says Dr. George L. Jacobson, Jr. of the University of Maine, there is virtually no time when the overall environment stays constant for very long. "That means that the configuration of the ecosystems is always changing," creating different landscapes in a particular area through geological time.[25]

In connection with change in relationships, Professor John Magnuson made a wonderful observation about foreseeing cause and effect. He said that all of us can sense change: the growing light at sunrise, the gathering wind before a thunderstorm, or the changing seasons. Some of us can see longer-term events and remember that there was more or less snow last winter compared with other winters or that spring seemed to come early this year. It is an unusual person, however, who can sense, with any degree of precision, the changes that occur over the decades of his or her life. At this scale of time, we tend to think of the world as static and typically underestimate the degree to which change has occurred. We are unable to sense slow changes directly and are even more limited in our abilities to interpret their relationships of cause and effect. Thus, the subtle processes that act quietly and unobtrusively over decades are hidden and reside in what Magnuson calls the "invisible present."[26]

It is the invisible present, writes Magnuson, that is the scale of time within which our responsibilities for our planet are most evident. "Within this time scale, ecosystems change during our lifetimes and the lifetimes of our children and our grandchildren."

It must be noted here that while it is possible to envision such serious accidents of human misjudgment as the meltdown at the nuclear plant in Chernobyl, Ukraine, or Iraq's invasion of Kuwait, the ultimate potential destruction of the planet with respect to human life will not be as apparent. Instead, it will occur slowly and silently, like the pollution of air, soil, and water—in the secret realm of the invisible present.

The world of Nature is a constantly changing dynamic set of relationships, and there are lags between the time when decisions are made and the time their consequences become evident. Such lags in time are the rule in ecological systems (including the ecology of all human societies); they separate cause and effect, which confuses our interpretation of Nature's world and makes it seem fickle and unsettled. Unless we adopt an attitude of far greater humility than we profess today, our lack of understanding of both events and processes in the invisible present will continue to be exceedingly costly to human society and its component communities.

It is thus essential that we become aware of our individual and social tendencies to suppress those changing relationships we deem uncomfortable and frightening. This will require a concerted, sustained effort at all levels of society to keep attention focused on the subtle changes in our environment that threaten our collective future.

For this reason, understanding and accepting responsibility for the following human relationships is central to sustainability: (1) *intra*personal, (2) *inter*personal, (3) between people and the environment, and (4) between people in the present and those of the future.

INTRAPERSONAL

An intrapersonal relationship exists within a person. It is the individual's sense of his or her own spirituality, self-worth, personal growth, internal power, and so on. In short, it is what makes that person conscious of and accountable for his or her own behavior and its consequences. The more spiritually conscious one is, the more other-centered one is, the more self-controlled one's behavior is, and the greater one's willingness to be personally accountable for the outcome of one's behavior with respect to the welfare of fellow citizens, present and future, and the Earth as a whole.

There is, however, a caveat to the above paragraph that most people overlook: to sustain your world, the community in which you live, and the family and friends you hold dear, you must first respect and honor your "self" as an individual and understand that the source of power in your world is within you. "Most people live, whether physically, intellectually, or morally, in a very restricted circle of their potential being," wrote William James. "Great emergencies and crises show us how much greater our vital resources are than we had supposed."

Yes, it would be grand to achieve national respect, even world acclaim, for ending injustice or hunger, for freeing the enslaved or finding the magic formula of public policy that would achieve true equality for all. But while focusing on the grand, one tends to neglect the basics—those "little details" of nurturing one's body, mind, and soul that often seem trivial and unrewarding when compared with changing the world.

Taking care of one's family and doing the work necessary to help it thrive may seem neither as important nor as pressing as serving the greater public good. But somewhere along the way to nobility, one's flesh is likely to become weak and one's family likely to disintegrate. And it is then one learns, if one is fortunate, that home is not only where the heart is but also were sustainability begins.

In working to achieve sustainable communities that are biologically, socially, and economically healthy, we must simultaneously work to develop individually healthy people, beginning with ourselves. We must learn, for example, that our source of power resides within us as internal motivation inspired to action, not in some external position of of authority.

People who succeed in changing their lives and growing beyond their present limitations do so because they not only know that success is the interpretation of an event and not the event itself but also because they recognize their own inner power, which gives them a sense of authority.

Authority is control—the right or power to command, enforce laws, exact obedience, determine, or judge. There are two kinds of authority embodied in this definition: that of a person and that of a position.

The authority of a person is an inner phenomenon, the inner source of power. It comes from one's belief in one's Self, that higher consciousness that acts as a guide in life when one listens to it. In contrast, a person who has the authority of position may have a socially accepted position of power over other people, but the sense of power can exist only if those people agree to submit their obedience to that person's position of authority. A person who holds a position of authority yet does not live from the authority within can only manage or rule as a dictator, through coercion and fear, but can never lead.

Leadership is an inner attitude, not an outer position or title. Leadership is possible only for one who has the inner sense of authority, regardless of his or her position, because true followership cannot be "managed" or "ruled." It can only be inspired. A leader can inspire a followership, because he or she has a vision that is other-centered, rather than self-centered. Such a vision springs from strength, those Universal Principles governing all life with justice and equity, not from the relatively weak foundation of insecurity and selfish desire. Leadership stems from inner authority, which gives the outer person a sense of authenticity. It is this authenticity that people respond to, and in responding, they validate their leader's authority.

Authenticity is the condition or quality of being trustworthy or genuine. It is the result of the harmony between what one thinks, says, and does and what one really feels—the motive in the deepest recesses of one's heart. The adage "deeds speak louder than words" is true as far as it goes, but what's left unsaid is that "motives speak louder than deeds." One is authentic only when one's motives, words, deeds, and attitude are in harmony.

Leaders inspire others by example to take responsibility for their own actions first, followed by the future welfare of their children and their neighbors. Although leaders are not the foundation of a sustainable community, they embody the spirit that motivates their neighbors and permeates a community with a drive to live within their highest moral values and within their ecological and economic means, thus ensuring, as far as possible, options for new leaders.

Attitude, be it one of leadership or otherwise, is the visible part of behavior, but motive is hidden from view. Ralph Waldo Emerson wrote, "Your attitude thunders so loudly I can't hear what you say," which in essence is saying that when one's visible behavior is out of harmony with one's projected motive, the attitude displayed thunders of a hidden agenda, an ulterior motive.

It is through understanding the authority of one's Higher Self and in feeling the authenticity thereof that one can allow oneself to be and to become. As you think, so you are, and as you become, so you attract. You are today what your thoughts have made you; you will be tomorrow what your thoughts will make you. And where you travel and who you are are determined by the choices, the decisions you make second to second, moment to moment.

We are all products of our choices, of our decisions—not victims of life. We make hundreds of decisions every day, and each decision represents a choice of roads Although most of us are not aware of all that our decisions have meant in shaping our lives, Robert Frost's poem strikes a universal chord:

> Two roads diverged in a wood, and I—
> I took the one less traveled,
> And that has made all the difference.

Each decision is a fork in our road of life; each fork is an alternative and a choice. The direction of our lives is a result of many little decisions; a few we remember, but most we don't because they are made unconsciously. We tend to remember the "big decisions," but we seldom realize that a single big decision is merely a collection of little decisions made along the way. We "give" just a little here and again a little there, and eventually we have positioned ourselves in entirely new directions.

Thus it is that we must take personal responsibility for our words and deeds, because the first step toward the sustainability of a community begins with us through the respect and the quality of care we give ourselves. Then we must extend that respect and care to our families, friends, and neighbors, because the sustainability of a community is but a reflection of the health of its citizenry as measured by how the people treat one another.

The upshot is that sustainability is possible to the degree to which people genuinely care about one another in a community or society. As individuals within our own communities, we need to search for those common values that, when acted upon in a communitarian fashion, can sustain the community, which brings us to interpersonal relationships.

INTERPERSONAL

Not enough can be said for cordiality, respect, and hospitality toward our partners in building community. If we use these basic human behaviors to frame our work together in striving for healthy, sustainable communities, we can focus more on the mutual values that bind us and less on the tension between our beliefs and attitudes. Here, an old saying is apt: Milk and honey are more easily swallowed than are sour grapes!

If, for example, someone is rushing blindly to get somewhere and shoves you out of the way, you have a choice in how you respond to being shoved. You can get angry, impatient, and say something nasty, or you can be patient, kind, and understanding, which brings to mind the comments of an old man knocked down by a rude young man in a hurry.

Upon rising from the floor, the old man regarded the younger for a moment and then said: "As you now are, so I once was. As I now am, so you shall one day be."

Our thoughts and actions are the seeds we sow each time we make a choice, and they will sooner or later come home to roost. If we do not like the outcome of our choice, we always have the choice of choosing again.

In fact, we always have a choice, and we must choose. We are not, therefore, victims of our circumstances but rather consequential products of our choices. And the more we are able to choose love and peace over fear and violence, the more we gain in wisdom and the more we live in harmony. This is true because what we choose to think about determines how we choose to act, and our thoughts and actions set up self-reinforcing feedback loops or self-fulfilling prophecies that become our individual and collective realities.

It is just such self-reinforcing behavioral feedback loops based on competition for resources that are destroying our environment. As long as competition is the overriding principle of our social/economic system, we can only destroy our environment because it has become the battlefield in which the war of competition is fought. Our overemphasis on competition in nearly everything fosters the material insecurity that often manifests as greed.

Another tendency of human beings is to defend a point of view when faced with a perceived threat to their sense of material survival. There are, however, as many points of view as there are people, and each person is indeed right from his or her vantage point. Therefore, no resolution is possible when each person is committed only to winning agreement with his or her position.

The alternative is to recognize that "right" versus "wrong" is a judgment about human values and is not a winnable argument, albeit each decision has a particular outcome that is more or less acceptable to a particular individual. It is best, therefore, to define the principle involved in the discussion as the fundamental issue and focus on it. An issue, usually perceived as a crisis, becomes a question to be answered, and in struggling toward the answer, both positive and negative options not only become apparent but also become a choice.

For us in the United States at least, crisis is too often in our point of view because we tend to perceive the world through a disaster mentality, regardless of evidence to the contrary. We tend to focus on and cling to a view of pending doom, in part because of the emotional discomfort of an unknown future heightened by daily news with graphic portrayals of disasters worldwide.

Fear, a projection into the future of unwanted possibilities, breeds weakness, a state in which there is little time or energy left to develop other areas of life. Out of the weakness of fear, men too easily and too often turn to war—or recently to terrorism—in an effort to assert what little power they think remains to them.

Recent years have demonstrated just how mindlessly cruel cultures can be when they live in proximity to and in dread of one another. The psychological and spiritual result of living under such heinous conditions deadens the mind while it renders savage the heart. Yet the cruelties within cultures and the violence of individuals are the essence of the news media.

Today's instantaneous news does not give us time to assimilate the stories within the context of global proportions. News came more slowly in olden times and could more readily be kept in perspective relative to the time and area it covered. Today, however, newsworthy disasters all seem to happen simultaneously and instantly in our homes via television and can become so overwhelming that we are emotionally numbed by them even as they augment our fear of our own unknowable future. In addition, insurance companies continually foster a disaster mentality.

Consider that insurance companies are betting, based on calculated probabilities, that nothing will happen to you as they take your money; you, by purchasing insurance, are betting blindly that a disaster will befall you in the future. You, therefore, are betting against yourself and your future. And it is just this disaster mentality that causes many frightened people to become increasingly self-centered.

We must therefore approach our mission of sustaining our communities within the context of sustainable landscapes by mutually supporting our common love of and need for a quality style of life. Too many community efforts stagnate and wither because people fail to nurture the social nature of human interaction.

Activities that build communities must include celebrations and create opportunities to simply have fun together. A community that works *and* plays together stays together and has an easier time forming partnerships to achieve a commonly held vision of a sustainable future. And just as sustainability begins with individuals, so too do the health and success of partnerships.

Forming partnerships must begin by resolving existing conflicts and clarifying commonly held values. Successful partnerships often start with a small group of people who develop (by consensus) a shared vision toward which to build. Partnerships are thus built on clear agreements among partners, which requires extensive and continuous communication, such as affirming understandings during and after meetings to focus the group and keep it focused on mutually desired outcomes and the steps necessary to achieve them.

It is critical, as partnerships evolve within a community, to create opportunities within and among them for success to be achieved. Even small victories in functional partnerships can strengthen and move the partnership forward to ever-more challenging tasks. It is important for partners to succeed together so they will want to stay at the table.

Partners must also be prepared for failure by learning how to "depersonalize" the problems that caused the failure. Learning how to become detached from outcomes is necessary in helping one another to overcome temporary setbacks. Pointing fingers and blaming not only waste time but also sap the energy needed to focus on success. Remember, anything worth doing well is worth doing badly for a while.

Strong partnerships require reciprocity, which is, after all, the essence of sustainability. The leadership must therefore pay particular attention to serving the needs and addressing the concerns of everyone in the group because people are more likely to give when they feel relatively assured of help in the future when they need it. This kind of caring and trust, which amounts to being one another's keepers, lends tremendous creativity toward resolving problems and accepting the inevitable setbacks.

Community partnership, like marriage, is a tough business. One must, with humility, make the commitment to take the time and have the patience to share power and resources. And one must be *committed to keeping his or her commitments,* which is a step most people either forget or do not understand.

Partnerships can only be created on a person-to-person basis. They must grow from the bottom up; they cannot be imposed from the top down. A partnership is an ongoing, ever-evolving process, not a fixed end point that once achieved is sufficient unto itself. The purpose of partnerships is to help people and communities empower themselves as they struggle toward social/environmental sustainability.

Some of the benefits derived from partnerships include: (1) helping to reduce costs by pooling available resources; (2) creating greater efficiency while protecting effectiveness by reducing overlapping and/or duplicative expenditures of time, energy, and finances; (3) creating more effective outcomes by addressing the interdependent, interactive complexities of an entire community; (4) having available many points of view focused on magnifying the innovation and creativity needed to resolve complex issues; (5) helping to build a sense of community with a sense of place that gives one a feeling of well-being and stability; and (6) increasing the social capital needed to make it through tough times over the long haul.

Effective partnerships require cooperation, coordination, and collaboration because our fortunes and fates are intertwined with those of others, which renders the concepts of self-sufficiency and isolated, individualistic activities increasingly meaningless. The interdependency of today's global society makes the world seem smaller while it commonizes the problems on an increasingly shared basis.

The greater the movement of a partnership from cooperation through coordination into collaboration, the more effective and better able it will be to deal successfully with complex issues. Cooperation requires partners to contribute resources, promote other people, and share information. Coordination demands the need to reconcile activities; tackle problems and deliver results more effectively; work jointly on projects, the results of which are beyond any one individual to achieve; and share in the cost of production. Collaboration requires people and organizations to share responsibility in managing resources and/or providing specific services by which new systems are created to accommodate a shared vision of the future and jointly made decisions.

Effective partnerships have members who respect one another. They listen well, refrain from dominating the discussion, focus on the problems or issues, and avoid making things personal. A good partner openly exposes his or her biases, which in turn creates an open, honest dialogue and develops a greater feeling of trust and sensitivity between and among partners.

When partners understand and appreciate one another's points of view and vulnerabilities, they can develop greater compassion for one another, which is manifested in mutual respect and personal humility. This, in turn, stimulates them to commit the time necessary to attend meetings, complete individual assignments, and stay involved throughout the process.

But there are also pitfalls along the road to partnership and sustainability. Some of these pitfalls include: (1) attending meetings but doing little else; (2) continuing conflicts among parties; (3) conflict over basic values, where no one is willing to change his or her position in order to negotiate; (4) confused purpose; (5) unrealistic vision and goals; (6) persistent self-centeredness; (7) abstractions never made concrete, such as specifying a piece of land; (8) interested parties not being included or refusing to participate; (9) inequitable distribution of power; (10) commitments of finances and time outweigh potential benefits; and (11) partnership is unnecessary because a single individual can achieve the desired outcome.

If all the pitfalls except self-centeredness are corrected, the partnership is still doomed. For a sustainable partnership to exist, self-centeredness must blend into other-centered teamwork. Setting aside egos and accepting points of view as negotiable differences while striving for the common good over the long term are necessary for teamwork. Unyielding self-centeredness represents a narrowness of thinking that prevents cooperation, coordination, collaboration, possibility thinking, and the resolution of issues. Teamwork demands the utmost personal discipline of a true democracy, which is the common denominator for lasting success in any social endeavor.

But even if we exercise personal discipline in dealing with current community/environmental problems, most of us have become so far removed from the land sustaining us that we no longer appreciate it as the embodiment of continuous processes. Attention is focused instead on a chosen product, the success or outcome of management efforts, and anything diverted to a different product is considered a failure. It is time, therefore, to reevaluate the philosophical underpinnings that gird our notions of Nature, partnership, community, and society and how they can be sustainably integrated into a common future.

BETWEEN PEOPLE AND THE ENVIRONMENT

Sustainability means that development programs must, to the extent possible, integrate the local people's requirements, desires, motivations, and identity in relation to the surrounding landscape. It also means that local people, those responsible for development initiatives and their effect on the immediate environment and the surrounding landscape, must participate equally and fully in all debates and discussions, from the local level to the national. Here, a basic principle is that programs must be founded on local requirements and cultural values in balance with those of the broader outside world, which includes understanding long-term ecological trends.

A trend is a line of general direction or movement, which in the environment is defined by a multitude of interacting factors, which include:

1. Location of event (e.g., on land, in water, in air, in the tropics, at the North or South Pole, in a valley)
2. Size of event (e.g., in someone's backyard, on an acre, over a landscape, over a continent)
3. Duration of event (e.g., ten seconds, an hour, a year, a century, a geological epoch)
4. Time of event (e.g., day, night, season, year)
5. Frequency of occurrence (e.g., once, hourly, daily, seasonally, annually)
6. Distance between events (e.g., an inch, a foot, a yard, a mile, 1000 miles, 10,000 miles)
7. Uniformity of event (e.g., uniform, roughly connected, disjunctive.)
8. Type of event (e.g., physical, biological, political, a combination)

The infinite variety of interactions among these factors creates an infinite variety of short-term trends that fit into a longer-term trend that fits into a still longer-term trend ad infinitum. Studying short-term trends (those that can be detected and perhaps understood) and projecting them over time may allow some degree of predictability of Nature's reactions to our decisions and actions.

There are two cautions, however. First, we must accept that all of these trends are ultimately cyclic and that their governing principles are neutral and impartial; the shorter the trend, therefore, the more imperative is our acceptance of Nature's neutrality and impartiality. Governing principles, whether biological or physical, are always neutral and impartial.

On the other hand, when we through politics assign values to Nature's actions based on our perceptions of "good" and "bad," we interject the artificial variable of partiality, which all too often clouds our vision. When this happens, we rob ourselves of our ability to predict the future with any degree of accuracy by rejecting Nature's impartial neutrality.

For example, a hillside meadow which bakes annually under the summer sun and turns brown is seen as having no value to a community because it does not produce a visible product the community wants. The meadow is therefore subdivided for a housing development. Within a year of completing the housing development, many of the community's wells begin drying up because the meadow had in fact been the water catchment that in secret supplied the wells.

Second, short-term trends must be viewed in relation to long-term trends and long-term trends in relation to even longer-term trends. The more we trace the present into the past, the better we understand the present. The more we project the present into the future, the more humble we need to be in our notion that we understand the present. A knowledge of the past tells us what the present is built on and what the future may be projected on. But this is true only if we accept past and present as a cumulative collection of our understanding of a few finite points along an infinite continuum—the trend of the future, which to many people is an abstraction.

Abstraction, according to physicist Fritjof Capra,[27] is a crucial feature of knowledge, because in order to compare and classify the immense variety of sizes, shapes, structures, and phenomena that surround us, we must select a few significant features to represent the incomprehensible milieu. We thus construct an intellectual map of reality in which things are reduced to their general outlines. This results in knowledge being a system of abstract concepts and symbols characterized by the linear, sequential structure that is typical of our thinking and speaking.

The natural world, on the other hand, is one of infinite variables and complexities, a multidimensional world without straight lines or completely regular shapes, where things do not happen in sequences but all at once, a world (as modern physics shows us) where even empty space is curved. It is clear that our abstract system of conceptual thinking is incapable of completely describing or understanding this reality and therefore is necessarily limited.

Thinking and knowledge in Western society have become so linear that we have forgotten that everything is defined by its relationship to everything else. Nothing exists by itself; everything exists in relation to something. In the example above, the community's well water existed in relationship to the hillside meadow and the precipitation that fell upon it. But the relationships were not understood until the housing development irreparably altered them.

Failing to account for a community's long-term supply of water in the face of short-term dollars to be made by a few people is dangerous, because changes in the spatial patterns of land use, which grossly alter habitats through time, may well be crucial to understanding the dynamics of landscapes and will have implications for many ecological processes. Changes in the patterns of landscapes may also be related to the flows of materials and energy across landscapes, such as the processes of erosion and the movement of water and sediments. Characterization of the relationships between changing patterns on the landscape and how those changes affect ecological processes is particularly important if we are to develop a more complete understanding of landscape dynamics, our effects on them, and their effects on our communities.

The reciprocal relationships between a community and its landscape may be summed up as a continuum of causes and effects that appear random in the short term and patterned in the long term. To understand this concept, we will look first at Silver City, New Mexico, and then at a community in northern Arizona to see how they affected themselves as communities by how they treated their immediate, respective landscapes.

New Mexico

The discovery of rich deposits of silver in 1870 marked the beginning of the mining industry in Silver City, New Mexico.[28] The area around the city was mined extensively and the forest was completely cut over between 1870 and 1887. The forest was decimated as fuel for steam boilers at mines, to build structures for mining, and to feed household fires.

In addition, the "grazing commons" around the city was used indiscriminately between 1870 and 1908 by cattle, sheep, goats, mules, burros, horses, and even swine in some places. During the dry years prior to 1900, according to old-timers, as many as 1,500 head of cattle would graze in close proximity to the city. Consequently, the grazing commons became badly overgrazed, and since most of the available forest had already been cut, there was practically no ground cover to hold the soil in place.

With vegetation gone and nothing to retain the water, the torrential rains that fell on the 21st of July 1895 left the people of Silver City with a painful reminder of Nature's awesome power. One day earlier, Main Street had been the city's primary north–south thoroughfare, the principal artery of commerce, the point of arrivals and departures, and the social center of town. Although the center of the street also provided moderate drainage, as it was two to three feet lower than either side, there was no indication that Nature had required it in the past.

Then the rains came. Pouring off the denuded water catchment, the large runoff created a monstrous gully out of Main Street, the bottom of which was 35 feet below the previous afternoon's street level. Subsequent floods, which climaxed in a two-day assault on the denuded water catchment in August 1903, scoured the gully down to bedrock, 55 feet below Main Street's original level of traffic, and the erosion continued some 15 miles south of town.

What happened in Silver City is a prime example of an irreversible result, in this case caused by indiscriminate mining, logging, and livestock grazing. Although the ecological damage to the city's landscape has been partially healed, it is not yet nearly as stable as it was prior to 1870. And as of 1980, a project was under way to establish a recreational walk along the sides of the gully known as the "Big Ditch."

Arizona

What happened in Silver City seems pretty clear because the evidence is documented in photographs and by the people who experienced it. Other cause-and-effect relationships between a community and its landscape are neither as dramatic in the short term nor as clearly obvious because they happen gradually over many decades. Consider, for example, a community in northern Arizona surrounded by a forest and dependent on its quality timber to feed local mills. What happens to such a community when the suppression of fire is introduced into the forest?

Remember, every ecosystem evolves inevitably toward a critical state in which a minor event sooner or later leads to a catastrophic event that alters the ecosystem in some way, such as a single spark that leads to a devastating wildfire. Consider, for example, that as a young forest grows old, it converts energy from the sun into living tissue, which ultimately dies and accumulates as organic debris on the forest floor. There, through decomposition, the organic debris releases the energy stored in its dead tissue. A forest is therefore a dissipative system in that energy acquired from the sun is dissipated gradually through decomposition or rapidly through fire.

Of course, rates of decomposition vary. A leaf rots quickly and releases its stored energy rapidly. Wood, on the other hand, rots much more slowly, often over centuries. As wood accumulates, so does energy stored in its fibers. Before suppression of forest fires in the early part of this century, forests burned frequently enough to generally control the amount of energy stored in accumulating dead wood by burning it up, thus protecting (fireproofing) a forest for decades, even centuries, from a catastrophic killing fire.

Eventually, however, a forest builds up enough dead wood to fuel a catastrophic fire. Once available, the dead wood needs only one or two very dry, hot years with lightning storms to ignite such a fire, which kills the forest and sets it back to the earliest developmental stage of grasses and herbs. From this early stage, a new forest again evolves toward old age, accumulating stored energy in dead wood and organizing itself toward the next critical state—a catastrophic fire—which starts the cycle over.

Therefore, a 700-year-old forest that burned could be replaced by another, albeit different, 700-year-old forest on the same acreage. In this way, despite a series of catastrophic fires, a forest ecosystem could remain a forest ecosystem. In this sense, the old-growth forests of western North America have been evolving from one catastrophic fire to the next, from one critical state to the next.

Although Gifford Pinchot, as first chief of the U.S. Forest Service, knew about fire, he was convinced it had no place in a "managed" forest. Fire was therefore to be vigorously extinguished, because conventional wisdom dictated that ground fires kept forests "understocked," and more trees could be grown and harvested without fire. In addition, surviving trees were often scarred by the fires, and this kind of injury allowed decay-causing fungi to enter the wood, thus reducing its quantity and quality. Finally, any wood not used for direct human benefit was considered an economic waste.

At this point, linear, commodity-oriented thinking entered the profession of forestry in the United States, and Pinchot's utilitarian conviction that fire had no place in a managed forest became both the mission and the metaphor of the young agency that he built. "Managed" in this sense came to mean any forested acre wherein someone perceived an economic interest in the trees. It was this notion of linearity—of economic waste if a potential commodity was not used by humans for the demonstrable benefit of humans—that so long ago introduced the "invisible present" of resource overexploitation and exhaustion into the profession of forestry and forest-dependent communities.

In Pinchot's time and place in history, he was correct and on the cutting edge, and the ecological problems caused by such thinking were unbeknownst to him. Nevertheless, incorporation of these ideas into forestry began to take their toll. Only now, decades after the instigation of fire suppression, has the significance of changes in forest composition, structure, and function and their influence on community sustainability become evident.

Recent evidence shows, for example, that some ponderosa pine forests in northern Arizona had only 23 huge, stately trees per acre in presettlement times. This presettlement density is in stark contrast to the current density of approximately 850 trees per acre, with predominantly small diameters.

The increase in density of trees is estimated to have caused: (1) a 92 percent drop in the production of grasses and herbs, which affects the grazing of livestock; (2) a 31 percent reduction in stream flow; (3) a 730 percent increase in accumulated fuels on the forest floor; (4) a 1700 percent increase in volume of saw timber; (5) a decrease from 115 to –8 in the index of scenic beauty; and (6) a habitat shift from open, savannah-like conditions in presettlement times to dense forest.[29]

This increase in the density of trees may also result in the decreased vigor and increased mortality of all trees, especially those of the oldest age classes. Finally, the increased closure of the canopy, the vertical continuity of fuel in the form of trees in the understory (a fuel ladder), and the high loads of fuel on the ground result in a severe hazard of crown fires (which burn through the tops of the trees high above the ground) and kill the forest. Such fires probably were exceedingly rare and localized in presettlement times, although ground fires (those that burn along the surface of the ground) were common before the settlers introduced grazing by domestic livestock and the Forest Service introduced the suppression of fire.

Only now, decades after Pinchot's instigation of the suppression of fire—one of the most prevalent forces of Nature—has the significance of changes in the composition and structure of forests in many areas become evident. During the last 80 to 100 years, since the advent of fire suppression, there has been a general increase in the number of trees and an increase in the amount of woody fuels on the forest floor. There also has been a decrease in the extent of quaking aspen (which often resprouts from roots following fire) and a corresponding increase in those species of trees that are more tolerant of the shaded conditions in closed-canopy forests. Some of these shade-tolerant trees have grown into the forest canopy and form a ladder up which a fire can burn from near the ground to the tops of large trees.

Although the role of fire in the physiological and ecological requirements of individual plant species may be relatively clear, there is greater difficulty in determining how fire regimes design whole forests. Most historical studies are hampered by effects of unknown events that can result in erroneous interpretations. It is therefore particularly important to study major ecological processes in an integrated fashion because such mechanisms are interactive and interdependent, and the *variability* in fire regimes is more likely to be important to plant communities than are the mean values computed from some arbitrary period of fire history.

For example, unusually long periods without fire may lead to establishment of fire-susceptible species. The simultaneous occurrence of such fire-free periods and wetter climatic conditions may also be extremely important to such species as ponderosa pine that have episodic patterns of regeneration (occur as specific, discrete episodes) as opposed to plants whose regeneration patterns are continual. Therefore, while statistical summaries of fire histories are useful in a general comparison of fire regimes in different forests, the influence of fire on a particular ecosystem is strongly historical. Some forests are more a product of unusual periods of climate and fire frequencies in the short term than they are of average or cumulative periods of climate and fire frequencies in the long term.

Although it is possible that climatic change could account for the increased number of large "wildfires," change in forest composition and structure is the most likely cause. Intensive study of historical fires has failed to document any cases of fire killing a forest by burning through treetops in the ponderosa pine forests of the American southwest prior to 1900.

In contrast, however, since 1950 numerous fires, exceeding 5,000 acres, have totally razed the forests down to mineral soil. The intensity of these fires is attributed to the amount of woody fuels on the forest floor and to dense stands of young trees within the forests—both of which have come about since 1900 and have brought more and more into question the sustainability of the supply of logs (even increasingly poorer quality logs) for the community's mills. Unless the community is committed to reversing the trend by reintroducing fire into the forest, the mills will ultimately close and the community will have to find another identity—all because an ecological process (fire) was eliminated from the ecosystem.

The only constant feature in a forest, as in life, is change. Each forest evolves in response to short- and long-term variations in its environment, such as climate, over which humans have no control.

Humans can neither arrest nor control changes wrought by Nature. If we could, the quivering balance through which Nature has produced the very forest we value and want to protect would be upset.

It now seems obvious that the effort to eliminate fire from forests is an economic rule made by humans. One major obstacle to changing this rule is our illusion of knowledge, which economic imagination draws with certainty and bold strokes, while scientific knowledge advances slowly by uncertain increments and contradictions.

One thing seems certain, however; the dynamic nature of evolving ecosystems, which are constantly organizing themselves from one critical state to another, precludes our ever being able to "manage" them. We can, however, treat an ecosystem in such a manner in the short term that we ensure, to the greatest degree possible, its ability to adapt to evolutionary change (such as global warming) in the long term in a way that may be favorable for us as local communities within the contexts of our landscapes.

A Community and Its Landscape

The setting of a community in the evolutionary sense helps define the community because people select a community for what it has to offer them within the context of its landscape. A logging community is therefore set within a context of forest, a ranching community within a context of lands for grazing, and a community of commercial fishers along a coastline, be it a lake or an ocean. The setting helps create many characteristics that are unique to the community. By the same token, the values and development practices of a community alter the characteristics of its surrounding environment.

In addition to the surrounding environment, the constructed environment within a community is also part of its setting and therefore its identity. Aesthetics, both internal and external to a community, is crucial to how the community defines itself through the philosophy it reflects in its livability. Much of what a community is saying about itself and how it cares or does not care about future generations is reflected in the physical structures with which it chooses to surround itself. This includes buildings, zoning, design of transportation systems, and the allowance of natural occurrences within the structured setting.

In turn, a community's world view defines its collective values, which determine how it treats its surrounding landscape. As the landscape is altered through wise use or through abuse, so are the community's ecological and social options altered in like measure. A community and its landscape are thus engaged in a mutual, self-reinforcing feedback loop of reciprocity as the means by which their processes reinforce themselves and one another.

Each community has physical, cultural, and political qualities that make it unique and more or less flexible. The degree of flexibility of these attributes in a community is important because sustainable systems must be ever flexible, adaptable, and creative. The process of sustainable development must therefore remain flexible, because what works in one community may not work in another or may work for different reasons.

Beyond this, the power of sustainable development comes from the local people as they move forward through a process of growing self-realization, self-definition, and self-determination. Such personal growth opens the community to its own evolution within the context of the people's sense of place, as opposed to coercive pressures applied from the outside.

Sustainable development encompasses any process that helps people meet their requirements, from self-worth to food on the table, while simultaneously creating a more ecologically and culturally sustainable and just society for the current generation and those that follow. Due to its flexibility and openness, sustainable community development is perhaps more capable than other forms of development of creating such outcomes because it integrates the requirements of a local community with those of the immediate environment and surrounding landscape, as well as neighboring communities. Sustainable community development can thus instill a relative balance between the local community and the bioregion within which it lies.

Bioregion

Bioregion refers to a geographical location and the ideas (collective consciousness) that have developed about how to live in that location with a sense of place. Within a given bioregion, the conditions that influence life are similar and thus have a similar influence on human occupancy. The notion here is that human cultures are differentiated at a bioregional scale in which the characteristics of the geographical region coincide with the collective consciousness of the people and are expressed as a specific culture.[30]

In terms of community sustainability, a bioregion must also be largely self-contained when it comes to an available supply of potable water. There is no chance of social/environmental sustainability without including the water catchment for the entire bioregion, because without a sustainable supply of water, sustainability is merely a paper exercise.

For a community to be socially and ecologically sustainable, it must simultaneously be as economically sustainable as possible, which means that communities must cooperate and coordinate within a well-defined bioregion. This seldom happens, however, because without a collective vision of sustainability within a well-defined bioregion, the communities are no more than economic colonies for national and international corporations.

The whole principle of colonialism is to exploit someone else's natural resources, shipping as much of the *principal* as possible, as fast as possible, to whichever market will pay the highest price. Thus, the more communities rely on outside markets, either for import or export of goods and services and/or jobs, the more they become economic and political colonies that progressively give up self-rule—and therefore democracy.

This means that a centralized national and international economy may be good for the corporate/political elite—but not for a local community. To be ecologically and socially sustainable, communities must learn to practice the *politics of place,* which is what bioregionalism is all about.

Bioregionalism is important because each community's economic sustainability demands that only the ecological *interest* of a bioregion is marketed. But the centralized corporate economy is in a constant feeding frenzy as it gobbles up all the ecological *principal* of all the available natural resources it can get. The legacy of this continual enrichment of the already wealthy minority is an increasingly fragile, ever-more endangered local environment.

Social/environmental sustainability is therefore dependent on a decentralized political/economic system of democracy if economic sustainability is to be achieved. Economic sustainability, in turn, is dependent on the cooperation and coordination of communities sharing a common vision of the greatest possible economic independence within the broad landscape of a well-defined bioregion.

Such economic independence—and with it the return of a free democracy—will not be easily wrested from corporate control. But it is possible, if communities can find the moral courage and political will to stand united within the umbrella of a shared vision of bioregional social/environmental sustainability for which they are willing to be accountable in the present—at least to the present generation and that of their children.

BETWEEN PEOPLE IN THE PRESENT AND THOSE OF THE FUTURE

We and our leaders must now address a moral question: Do those living today owe anything to the future? If our answer is no, then we surely are on course, because we are consuming resources and polluting the Earth as if there were no tomorrow. If, on the other hand, the answer is, "Yes, we have an obligation to the future," then we must determine what and how much we owe, because our present, nonsustainable course is rapidly destroying the environmental options for generations to come. Meeting this obligation will require a renewed sense of morality—to be other-centered in doing unto those to come as we wish those before us had done unto us.

To change anything, we must reach beyond where we are, beyond where we feel safe. We must dare to move ahead, even if we do not fully understand where we are going, because we will never have perfect knowledge. We must ask innovative, other-centered, future-oriented questions in order to make necessary changes for the better.

True progress toward an ecologically sound environment and an equitable world society will be expensive in both money and effort. The longer we wait, however, the more disastrous becomes the environmental condition and the more expensive and difficult become the necessary social changes.

No biological shortcuts, technological quick fixes, or political promises can mend what is broken. Dramatic, fundamental change, both frightening and painful, is necessary if we are really committed to the children of the world, present and future. It is not a physical question of whether we *can* change, but rather one of human morality—whether we *will* change. Whatever our decision, we, the adults of the world, bequeath to the children of the world the consequences thereof through our democratic system of government for which we are *all* responsible, whether we participate or not.

SOCIAL GOVERNANCE 9

There are a number of different ways to govern, and each is more or less unique. Colonialism, for example, is the situation in which one country extends and/or maintains its control over another. Colonialism is openly exploitive. Its purpose is to take, not to give. Communism, on the other hand, is a social system ostensibly characterized by the absence of classes and by common ownership of the means of production and subsistence. Dictatorship is a form of government ruled by someone who has absolute authority and supreme jurisdiction over the people, especially someone who is tyrannical or oppressive. A monarchy is similar in definition to a dictatorship, with the exception that acknowledged hereditary familial lines often determine who is king, queen, emperor, or empress. And finally, there is democracy, which in its best form is a participatory government of the people, by the people, and for the people.

DEMOCRACY AS THE CONTEXT FOR SOCIAL RELATIONSHIPS

Democracy is the backbone of local community development and ultimately community sustainability, and, like sustainable community development, governance is an evolutionary process. And because every phase of organizational growth and development within a community is determined by its membership, so too must the appropriate forms of democratic governance (structure/process) be determined by the membership.

It is thus critical to understand something about democracy as a practical concept. Democracy is a system of shared power with checks and balances, a system in which individuals can affect the outcome of political decisions.

Within this system, we can make real and profound changes. All that is required is education, a purposeful vision, and political will, which means having the courage to come together and to choose to act, the knowledge to know how to act, and the vision to know where to go.

"I know no safe depository of the ultimate powers of the society but the people themselves," penned Thomas Jefferson in a letter to William Charles Jarvis on September 28, 1820, "and if we think them not enlightened enough to exercise their control with wholesome discretion, the remedy is not to take it from them, but to inform their discretion by education."

"Enlighten the people generally," wrote Jefferson, "and tyranny and oppression of body and mind will vanish like evil spirits at the dawn of day." Today's education is not enlightening the people generally, however. While economist E.F. Schumacher shared Jefferson's notion about education, he also saw its folly in contemporary industrial societies: "If Western civilization is in a state of permanent crisis, it is not far-fetched to suggest that there may be something wrong with its education....More education can help us only if it produces more wisdom."

Democracy is designed to protect individual freedoms within socially acceptable relationships with other people individually and collectively. People practice democracy by managing social processes themselves. Democracy is another word for the responsibility of self-directed social evolution, which might be what William James was referring to when he stated: "I will act as if what I do makes a difference."

Democracy in the United States is built on the concept of inner truth, which in practice is a tenuous balance between spirituality and materialism, intuition and knowledge. One such truth is the notion of human equality, in which all people are pledged to defend the rights of each person, and each person is pledged to defend the rights of all people. In practice, however, the whole endeavors to protect the rights of the individual, while the individual is pledged to obey the *will* of the majority, which may or may not be just to each person because the majority may not be just. And when the majority is unjust, it should not win.

Democracy is therefore "tough love," because one must think about and be accountable for one's own conscience and behavior. Democracy is tough love because one must constantly focus on the principle above and beyond the individual. Democracy is tough love because it is time consuming, expensive, sloppy socially, cumbersome, and inefficient, but these are the very qualities that create and maintain the greatest freedom possible in a diverse society. Be that as it may, "There is nothing harder than the softness of indifference," wrote Ecuadorian essayist Juan Montalvo, and time has proven him correct.

In fact, our democracy was designed as tough love, to be inefficient through the separation of powers, which not only minimizes the concentration of power but also stresses an old maxim: A government that governs least governs best. That is why the original intent of the Constitution was for states and municipalities to have more power for self-government than they do today, and it is precisely this local power that is required for sustainable community development. There is a caveat to this statement, however; for one's community to be sustainable and our democracy to be lasting, we must individually and freely be willing to recognize and abide by the common good in our decision making.

"Every age and generation," wrote Thomas Paine, "must be as free to act for itself in all cases as the ages and generations which preceded it. The vanity and presumption of governing beyond the grave is the most ridiculous and insolent of all tyrannies." In a similar tone, Thomas Jefferson concluded that "no society can make a perpetual constitution, or even a perpetual law. The earth belongs always to the living generation."[31]

Although Paine's and Jefferson's comments may sound like each living generation has the right to use the world in which it lives with total abandon and without a thought for future generations, such is not true. Paine's first statement ("Every age and generation must be as free to act…as the ages and generations which preceded it.") says in effect that each generation must save and pass forward all possible options to the next generation or the latter will not be as free to choose as the preceding one.

Paine's second statement ("The vanity and presumption of governing beyond the grave is the most ridiculous and insolent of all tyrannies.") points out that we cannot force on any future generation our current set of values and that to try is both futile and arrogant. Beyond that, his statement smacks hard the face of industry, which all too often says that unless there is a guarantee that the next generation will follow our current set of values, there is no point in saving resources for it. Industry therefore quickly exercises its option of being in the present and liquidates the coveted resources before the future has a voice.

Jefferson's first statement ("No society can make a perpetual constitution, or even a perpetual law.") is similar to Paine's second. On the other hand, Jefferson's second statement ("The earth belongs always to the living generation.") states clearly that viable governance must always change to meet the times, which means people must change as the times change, something we are loath to do. And change comes first to local communities and their environments.

Because change comes first to local communities, we must be involved in local decision making (city and county). We cannot afford to turn our personal responsibilities over to absentee state and federal governments and allow them to be the dominant mechanisms of social control and regulation. Family and spiritual values and public education are supposed to play a mediating role in our society, which requires personal maturation on the part of physical adults, but we lose the balance of maturity when we abnegate our communal responsibilities.

Democracy is tough love because it requires maturity, which means restraining our self-centeredness and recognizing and accepting that *we are not the center of the world,* that we cannot with impunity lay waste to other countries and the future for the sake of our insatiable material appetites. Maturity means we must stop being primarily acquisitive and begin concentrating on moral, spiritual, and cultural development, the fruit of which is real equality among people—among all people of all countries, between genders, and among generations. Maturity means acting as true adults, rather than spoiled, self-centered children in adult bodies. Whether we mature as individuals or not is our choice, a choice with a price, as Thomas Jefferson alluded when he warned: "People get the government they deserve."

Thus, like marriage, democracy is tough love because we must stay together and struggle together toward a perfection of government we shall never achieve. But if we drift apart, we divorce from the principles of democracy, and dissolution of our form of society becomes inevitable.

It is this whole notion of true equality within diversity that brings up the "will" of the majority in terms of freedom in a democracy. Nothing that we know of is totally free; rather, everything expresses freedom within some sort of limits. Just as there is no such thing as a truly "free market" or an "independent ecological variable," so are individual and social autonomy protected by moral limits on the freedom within which individuals and society can act. To this effect, author

Anna Lemkow[32] lists four propositions of freedom: "(1) An individual must win freedom of will by self-effort, (2) freedom is inseparable from necessity or inner order, (3) freedom always involves a sense of unity with others beyond differences, (4) freedom is inseparable from truth—or, put the other way around, truth serves to make us free."

Lemkow goes on to say:

> We tend to think of freedom as dependent on circumstantial or external factors, but these propositions point us inward, suggesting rather that freedom is a state of consciousness and...depends on ourselves. Indeed it is something to be won, something to be attained commensurately with becoming more truthful, or more attuned to and aligned with the abiding inner, metaphysical, or moral order or law.
>
> Socio-political and economic freedom or liberty, in turn, would depend (at least in the longer term) on the predominant level of consciousness of the citizenry.

Lemkow is positing that a human being is not completely free to begin with, but possesses the potential capability of self-transformation in the direction of fuller freedom. Freedom is not free, however; it demands sacrifice not only on the field of battle but also in the conference room and in the home.

There will never be a free democracy or a free world until it is composed of people who free themselves. A real democracy cannot be made of slaves, whether intellectual, emotional, economic, or physical. A person is a slave until he or she recognizes and accepts immediate personal responsibility for his or her character, behavior, and destiny.

The freedom of democracy is hard won, difficult to retain, and easily lost. It requires nurturing by a vibrant, creative, civic society working outside the established administration of the government because democracy is at best a risky, open-ended experiment that requires every generation to face a new set of responsibilities and challenges to match its times.

To the degree that the responsibilities and challenges are not met, society will sink into a deficit of virtue in like measure. This in turn will lead to a corresponding spiritual bankruptcy, which, should it be deep enough, will steal freedom and hope from the human soul and sound the death knell of democracy.

To flourish, democracy requires respect for others and excitement in the exchange of ideas. People must learn to listen to one another's ideas, not as points of combative attacks but as different and valid experiences in a collective reality. While they must learn to agree to disagree at times, they must also learn to accept that, like the proverbial blind people feeling the different parts of an elephant, each person is initially limited by his or her own perspective. When these things happen, people are engaged in the most fundamental aspects of democracy and come to conclusions and make decisions through participative talking, listening, understanding, compromising, agreeing, and *keeping* their agreements in an honorable way.

In a democracy, *connection* and *sharing* are central to its viability because a democracy only works when it is being practiced. And a democracy can only be practiced by an educated public. A sound education is therefore an absolute necessity for the survival of democracy. Withhold education, and dictatorship is a virtual certainty.

In a democracy, people are *not* required to separate feelings from thoughts concerning a topic. Their roles—as teachers, students, leaders, facilitators, and followers—fluctuate within and across the issues. The importance of a democratic system lies in its connection to people's lives, its relevancy to their own experiences, and the real problems and issues they daily face. Practicing democracy can be thought of as education in and for life.

The challenge, therefore, is to engage people in the democratic process, which is difficult when they confuse the government with the administration of the government or when the administration becomes so dysfunctional that people despair in their seeming inability to fix it. To this end, Thomas Jefferson made the following observations concerning the federal government of the United States:

> When all government, domestic and foreign, in little as in great things, shall be drawn to Washington as the center of all power, it will render powerless the check provided of one government on another, and will become as venal and oppressive as the government from which we separated [1821].[33]

> If ever this vast country is brought under a single government, it will be one of the most extensive corruption, indifferent and incapable of a wholesome care over so wide a spread of surface. This will not be borne, and you will have to choose between reformation and revolution. If I know the spirit of this country, then one or the other is inevitable. Before the canker is become inveterate, before its venom has reached so much of the body politic as to get beyond control, remedy should be applied [1822].[33]

Although we know of no perfect government, a just government must be founded on truth—not knowledge, something we in the United States have all too swiftly forgotten. To achieve such government, it must be based on service (where people are other-serving) rather than on power (where people are self-serving). Therefore, development is only sustainable and environmental protection is only possible if the government is accountable to its people beyond monied special interest groups and political lobbyists.

We live in an increasingly complex society of intense competition and materialism. In such a society, people commit evil acts in order to gain power, both through positions of authority and financial success. People commit evil acts while falsely expecting to benefit by them, thinking that such benefits will somehow bring happiness. But in the end, as Socrates warned, the guilt of the soul outweighs the supposed material gains. Thus, because people lack perfect knowledge and perfect motives, democracy must be continually practiced and continually improved through that practice.

Nevertheless, the *people* are the government, but they can govern only as long as they elect to use the constitutional system for empowerment at the local community level. It is important to understand that empowerment is personal self-motivation. No one can empower another; one can only empower oneself. One can, however, give others the psychological space, permission, and skills necessary to empower themselves and then support their empowerment. Beyond that, one can help in the process of empowerment and can increase the chances of success by recognizing another's accomplishments each step of the way. That is true democracy.

Therefore, in the case of the United States, where the government is of the people, by the people, and for the people, when the people empower themselves, they are the government, and it is the administration of that government (and not the government itself) that resides in city hall, the county seat, the state capital, or Washington, D.C. The administration becomes the government when the people turn their power over to the administration and in effect say, "I am a victim and cannot change the system" or "Take care of me."

When this happens, democracy is endangered because, as Carl Jung noted, "to turn the individual into a function of the State, his dependence on anything beside the State must be taken from him [or voluntarily given up]." Professor Alexander Tayler described this process in a democracy over 200 years ago, when what is now the United States was still a British colony. Tayler penned the following:

> A democracy cannot exist as a permanent form of government. It can only exist until voters discover that they can vote themselves largesse from the public treasury. From that moment on, the majority always votes for the candidates promising the most benefits from the treasury, with the results that a democracy always collapses over loose fiscal policy, always followed by a dictatorship.
>
> The average age of the world's greatest civilizations has been two hundred years. These nations have progressed through this sequence:
>
> from bondage to spiritual faith;
> from spiritual faith to great courage;
> from great courage to liberty;
> from liberty to abundance;
> from abundance to selfishness;
> from selfishness to complacency;
> from complacency to apathy;
> from apathy to dependence;
> from dependence back again into bondage.[34]

If what Tayler says is true, then we may well be on the down side of the democratic cycle, a thought echoed in 1993 by William Bennett, former secretary of education: "It's a misdiagnosis to say [America's] problem is economics. It's a cultural decline. Our problems are moral, spiritual, philosophical, and behavioral."

But that does not mean that the cycle is irreversible. The cycle can be altered for the better by conscious choice. Democracy can be a viable system by inviting a constant reinterpretation of itself based on asking morally, socially, environmentally, and economically sound, farsighted questions. In this way, it is always in a state of becoming, which continually interweaves it within the intimacy of life.

For democracy to remain viable, its principles and processes must be used, because they form an interactive, interdependent system of balancing and integrating data as testable knowledge and intuition as untestable human truth from which we derive a kaleidoscope of ever-changing perceptions. A working democracy is thus predicated on finding the point of balance through compromise in such a way that the rifts between opposites can be minimized and healed while protecting the integrity of Nature, from which comes our source of physical energy.

Understanding the fundamental processes of a free democracy is critical if people are to see the value of their participation in making the democracy work, because, as previously stated, the principles of democracy only function when democratic processes are actually available and used. The basis of these processes allows and encourages people to give their obedience to those concepts or principles whose ethical values they hold dear and to withhold their obedience from those with which they disagree.

Consider, for example, that all we have to offer the future is options (which are choices to be made), and those options, both biological and legal, are held within the environment (including local communities) as a living trust, of which we are the legal caretakers or trustees. Although the concept of a trustee or a trusteeship seems fairly simple, the concept of a trust is more complex because it embodies more than one connotation.

A living trust, for instance, is a present transfer into trust of property, including legal title, whether real property or personal property, such as livestock, interest in a business, or other property rights. The person who creates the trust can watch it in operation, determine whether it fully satisfies his or her expectations, and, if not, revoke or amend it.

A living trust also allows for delegating administrative authority of the trust to a professional trustee, which is desirable for those who wish to divest themselves of managerial responsibilities. The person or persons who ultimately benefit from the trust are the beneficiaries.

The environment (and its local communities) is a "living trust" for the future. A living trust, whether in the sense of a legal document or a living entity entrusted to the present for the future, represents a dynamic process. Human beings inherited the original living trust—the environment—before legal documents were invented. The Earth as a living organism is the living trust of which we are the trustees and for which we are *all* responsible.

Throughout history, administration of our responsibility for the Earth as a living trust has been progressively delegated to professional trustees in the form of elected officials. The same is true for local community, county, state, and national government. In so doing, we empower elected officials with our trust (another connotation of the word, which means we have firm reliance, belief, or faith in the integrity, ability, and character of the elected official who is being empowered).

Such empowerment carries with it certain ethical mandates, which in themselves are the seeds of the trust in all of its senses, legal, living, and personal:

1. "We the people" own the trust and the elected officials and their professionals act as our appointed trustees.

2. We have entrusted our elected officials to follow both the letter and the spirit of the law in the highest sense possible.
3. We have entrusted the care of the environment and our local communities to elected officials, all of whom have sworn to accept and uphold their responsibilities and to act as professional trustees in our behalf.
4. We have entrusted to these officials and professionals the livelihood and health of our environment and community. Through the care of these officials and professionals, they are to remain living, healthy, and capable of benefiting both present and future generations—the beneficiaries.
5. Because we entrusted the environment and our community as a unified "present transfer" in the legal sense, we have the right to either revoke or amend the trust (the empowerment) if the trustees do not fulfill their mandates.
6. To revoke or amend the empowerment of our delegated trustees if they do not fulfill their mandates is both our legal right and our moral obligation as hereditary trustees of the Earth and as voluntary citizens of our community.
7. Apart from elected officials whom we have designated to act as trustees for us in the present, we, as adults, are the hereditary trustees of the options belonging to the generations of the future—a trusteeship from which we cannot divorce ourselves.

How might this work if we are both beneficiaries of the past and trustees of the future? To answer this question, we must first assume that our elected officials and their professionals are both functional and responsible. *The ultimate inheritance entrusted to the present generation for all those of the future would be to pass forward the capital of the trust, with as many of the existing options intact as possible.*

These options would be forwarded to the next generation (in which each individual is a beneficiary who becomes a trustee) to protect and pass forward in turn to yet the next generation (the beneficiaries who become the trustees) and so on. In this way, the maximum array of biological and cultural options could be passed forward in perpetuity—the essence of sustainability.

If, however, the elected officials and professionals did not fulfill their obligations as trustees to our satisfaction, then their behavior could be critiqued through the electoral process and/or the judicial system, assuming that the judicial system is both functional and responsible. Our children, when they become of voting age, become our judges and can critique our trusteeship of their future through the electoral process and/or the judicial system, as we did our predecessors.

The invisible present embodied in our decisions as trustees of today can create a brighter, more sustainable vision for the generations to come, who are the beneficiaries of the future when they stand in their today. In order for this to happen, however, we must actively participate in the democratic rule of our communities so that they become as sustainable as possible in partnership with their surrounding environments within a bioregion. We must understand and accept that a sustainable community is, in a sense, the institution in which the living trust is housed and protected. We must also make our judicial system just and responsible to all generations, something we have not yet chosen to do. And it is, after all, only a choice, which is the very foundation of democracy.

Democracy sets up and maintains the information/political feedback loops through which human values, intuition, information, and cultural innovations are funneled into the governmental system that today drives our communities.

THE ROLE OF DEMOCRATIC GOVERNMENT

Although the critical role of developing sustainable communities must be centered among the members of local communities themselves, all levels of government can and must play a quality role in nurturing the creation, process, and maturation of partnerships. In fact, a point often forgotten by parochial interests is that people who work for governments may also be members of a local community and care just as much about its future as do citizen activists.

Community partnerships do well when they include public servants who have special access to resources, can move the processes of government, and can lend expertise to efforts of inventing or reinventing the relationship between private citizens and government. We, as citizens of communities, must learn to cherish those leaders in government who understand that if community partnerships are given the correct tools with which to create their own sustainable futures, the work of government will be not only less intrusive but also more effective.

In short, communities and government policymakers need to work together to change both current laws and government-to-citizen relationships so that local communities can participate meaningfully in caring for their surrounding landscapes for their own benefit as well as the benefit of those beyond their parochial borders. But any such activity must fit—first and foremost—into the biologically sustainable capacity of the landscape, which generations of the future must inherit as we adults leave it for them.

In this spirit, private landowners must be encouraged (even if they respond only to the tenets of Rational Economic Man through tax incentives, revised inheritance laws, and technical assistance) to participate in cross-ownership, results-oriented ecosystem care. After all, we humans—with the power to reason, the power to plan, and a responsibility to all future generations—are the trustees for the future. As such, it falls on us to renew the environment and revive the essence of community, which takes leadership.

Provide Exemplary Leadership

Governments cannot empower people, but they can give them the proper tools and the capacity to work actively to achieve the outcomes they want. For instance, incentives for good trusteeship of a local community's natural resources by the community itself can replace tax breaks used to recruit multinational corporations with little or no stake in either the ecological sustainability of a community's natural resources or the high-quality jobs needed to sustain a community's cultural well-being.

In addition, organizations for economic development, rather than merely ladling out grants, can provide expertise and financial backing so that local community partnerships can learn how to create their own vision and plan strategically how to achieve their goals. Such support would include assistance in strategic planning for businesses, including marketing, that will help communities retain those small businesses deemed critical to their social sustainability.

Governmental leadership can also employ measures of performance for agencies and their staffs to reward those actions that help the growth and sustainability of community partnerships. In this spirit, a governor can require state agencies to use cross-agency decision making and can combine funding for agency programs in order to increase the efficient and effective access of local communities to resources at the government's disposal. And greater devolution of decision-making authority to the local level can make participation in community partnerships more effective, which creates the opportunity for community self-empowerment.

A prerequisite for sustainable development in a local community is that it must be inclusive, relating all relevant disciplines and special professions from all walks of life. Setting a good example is one of the most important functions of any local government involved in implementing the principles and practices of sustainable community development. Leading by example—breaking down bureaucratic barriers through interdisciplinary crossing of departmental lines, recycling and buying recycled goods, car pooling, and providing day care and flexible working hours—increases not only the capacity of a government to govern but also its effectiveness *and* efficiency.

It is thus important for governments to both identify departmental and community links concerning mutually interrelated issues and to bring all people affected to the table in an effort to collectively resolve shared problems, which means dealing with human diversity. Understanding and accepting diversity allows us to acknowledge that each of us has a need to be needed, to contribute in a meaningful way. It also enables us to begin admitting that we do not and cannot know or do everything and that we must rely on the strengths of others with complete trust.

Diversity of thought, culture, expertise, and economic status thus allows all persons to contribute to the development process in a special way, making each individual's unique gift a part of the effort necessary to create a sustainable local community. Accepting diversity helps us to understand the need each person has for equality, identity, and opportunity in the process. Recognizing diversity gives us all a chance to provide meaning, fulfillment, purpose, and the gift of our talents to our community and future generations. (Conversely, just as simplifying an ecosystem increases its vulnerability to destruction, so too will segregating diverse elements within a community lead to its social, moral, and economic decay, which may then spread throughout society, one community at a time.)

Assuming people accept the notion of diversity, what is it they most want from the development process? People want the most effective, productive, and rewarding way of working together to achieve a common end. They want the process and the relationships forged therein to meet their personal needs for belonging, meaningful contribution, the opportunity to make a commitment to a special place—their community, the opportunity for personal growth, and the ability to exert reasonable control over their destinies.

Increasing control over personal destinies and thus the destiny of a community can be increased if federal agencies will focus on a community's goals for social/environmental sustainability. Having said this, however, it must be recognized that it is federal agencies (not local communities) that have jurisdiction over *public* lands and are thus ultimately responsible for the ecological sustainability of those lands as a national legacy for the generations to come.

It is also possible to create government-to-community trusteeship contracts, which recognize that it is the results that count, not necessarily how many rules were followed. Such action can devolve authority closer to the citizens of a community and simultaneously allow employees at all levels of government to empower themselves to achieve quality results, which may well improve citizen participation in local government.

Improve Citizen Participation

A vitally important component of sustainable community development is local citizen participation in planning, implementing, and monitoring programs, policies, and projects. The goal is to improve the quality of popular participation instead of merely its quantity.

Sustainable community development is based on the assumption that the best ideas usually come from the people, not the policymakers. Therefore, active participation in a local community is necessary to direct the process, which means, for example, taking part in citizen administrative boards and in town hall meetings and through local grass-roots activities.

As a process, sustainable community development exposes citizens to the ramifications of their thoughts and actions on others, their local environment, and the surrounding landscape, as well as motivating and organizing people to direct change within the context of a shared vision for their collective future. Its aim is for citizens to control the developmental process by feeding ideas and information to the governing body through self-empowered organizations.

People want the most effective development process possible, one that is honestly used through participation in a truly democratic way. Participative development must begin with a firm belief in the potential of people. It arises both out of a leader's heart and his or her personal commitment to people and out of the heart of the democratic principle: the right to an open, accessible process; the right and duty to influence decision making; the right and duty to understand the results; and the duty to be accountable for those results.

To accomplish participative development, leaders must create and maintain emotionally safe environments within which people can develop quality relationships with one another. Creating such an environment requires at least six things: (1) respect for one another; (2) understanding and accepting that what people believe precedes policy and practice; (3) agreement on the rights of participation in and access to the planning process; (4) understanding that most people work as volunteers and need personal covenants, not legal contracts; (5) understanding that relationships count more than structure because people—not structures—build trust; and (6) protecting the process against capture by self-serving financial interests.

The development committee's needs are best met by meeting the needs of its individuals. If this is done, development can be productive, rewarding, meaningful, maturing, enriching, fulfilling, healing, and joyful. Participative development is one of the greatest privileges in our democracy and one of our greatest responsibilities.

Nevertheless, the creative development process is difficult to handle because in such a process almost everyone, at different times and in various ways, plays four roles: one as creator, another as implementer, a third as temporary leader with a specific expertise demanded by a given circumstance, and finally as follower, supporter, and helper.

Although implementation is often as creative as the questions to which it is responding, it is at this very point that leaders and managers may find it most difficult to be open to the influence of others. Nevertheless, by conceiving a shared vision and pursing it together, a local community's problems of cultural adaptability and sustainable development can be resolved, and the community members may simultaneously and fundamentally alter their concept of adaptability, sustainability, and development. But this requires "joint ownership" of the development process.

The heart of sustainable development is joint ownership of the process for each person involved. Because owners cannot walk away from their concerns, everyone's accountability begins to change. Ownership demands increasing maturity on everyone's part, which is probably best expressed in a continually rising level of literacy: participative literacy, ownership literacy, sustainable development literacy, and so on. And ownership demands a commitment to be as informed as possible about the whole.

Joint ownership is an intimate, personal experience in that each person commits himself or herself not only to the process but also to the outcome. One's beliefs are connected to the intimacy of one's experience and come before and have primacy over policies, standards, or practices. This intimate, personal commitment to the development process affects one's accountability and draws out one's personal authenticity.

No development process can amount to anything without the people who make it what it is. It is initially what the people are and finally what the people become. People do not grow by knowing all the answers; they grow by living with the questions and their possibilities. The art of working together thus lies in how people deal with change, how they deal with conflict, and how they reach their potential.

The intimacy of ownership arises from translating personal and community values into a plan for a sustainable future that seeks its excellence in a search for truth, wisdom, justice, and knowledge—all tempered by intuition, compassion, and mercy. The people of a community must therefore make a covenant, a promise with one another, to honor and protect the sacred nature of their relationships so that each may reflect unity, grace, poise, creativity, and justice. If they base decisions on the intrinsic value of human diversity, and if they base decisions on the notion that every person brings a unique offering to the development process, then inclusivity will be the only path open to them.

Including people—really including people—in the development process means helping them to understand the process, their place within it, and their accountability for the outcome. It means giving others the chance to do their best according to the diversity of their gifts, which is fundamental to the equality that environmental justice requires and democracy inspires. Finally, a community must be committed to using wisely and responsibly its environment and its finite resources, which means a conscious, sustainable, reciprocal relationship between the local community and its surrounding landscape.

To create the desired change, however, it is essential that all affected groups in the community be involved in the process and trained in the skills of leadership. It is further necessary that the people responsible for a local program, policy, or project be involved in its creation and monitoring to increase the probability of a successful outcome. If they are not involved throughout the whole process, a political problem arises because sustainable community development is initially site specific, and that, in our experience, inevitably brings up either turf struggles or a blatant denial of both responsibility and accountability by passing the buck.

Top-down planning does not work because communities have no vested interest in doing what they feel will benefit other communities at their own perceived momentary expense. Thus, little gets done with sufficient forethought to be of real long-term social/environmental benefit to the future of the community. Oh yes, there are the interminable meetings, but little significant action and even less accountability.

As long as the majority of the people in a community, county, state, or nation are predominantly self-centered, and thus myopic, each and every level of government must see a clear—and often immediately personal—advantage before cooperation and coordination become a reality. This is important, because to cooperate and coordinate implies the willing acceptance of both responsibility and accountability, which most people avoid whenever possible. Whatever people do, they do by choice, whether or not they understand it or accept it.

Protect and Enhance People's Choices

Choice—a simple but powerful word—changes the world every day in ways both great and small. People, having appropriated choice from the dawning of humanity, value it, vie for it, and die for it. Yet many people have little or no concept of the sense of freedom that comes with the ability to choose or the sense of responsibility that comes with the consequences of one's choice.

Thurgood Marshall and Mahatma Gandhi understood it. They made no apologies for wanting it equally for themselves and for all other people. They did much to make the world see that individual choice—implied by basic human rights—has no meaning unless it is universally available to all people. The ability to choose confers upon each individual a sense of value, self-confidence, and the dignity of being human.

But how does one select among the billions of choices? Are we so addicted to and paralyzed by our myriad choices that we as a society are losing our ability to choose collectively when the need to do so is more urgent than ever? Are we trying desperately to see life as a series of painless options by ignoring the hard choices about the future health of the environment and thus the health of our children, their children, and their children's children? Or are we purposefully passing the buck into the future, so those for whom we choose but give no voice can pay the social/environmental bill when it comes due?

If the outcome of a hard choice is good, then people will scarcely remember the pain. The question is whether we have the wisdom and courage to be responsible and not only make the hard choices but also live with their consequences.

Because choice is the essence of not only democracy but also sustainable community development, no greater disservice can be rendered by those in power than to unjustly limit the people's power to freely choose. And limiting the people's choice in favor of the traditional short-term economic desires of two special interest groups is in essence what the Oregon legislature has done.

> A 1995 land-use law could make it harder for the public to comment on residential development requests.
>
> House Bill 3065, pushed through just days before the end of last year's Legislature, cuts the approval process in half and forces comments to focus on a subdivision's legal merit.
>
> "The idea is to get emotions out of land-use decisions," said Drake Butch, of the Home Builders Association of Metropolitan Portland. The lobbying group, along with the Oregon State Home Builders Association, pushed for the bill's passage.
>
> "It's not fair to developers to keep delaying the process with appeals when their proposed developments meet all the legal requirements," Butch said.
>
> He said it's costing developers too much money to go through the approval process because of delays for public comment and appeals.[35]

The effect of this law is to (1) steal choice and self-determining government from the people who live in the area of the proposed development; (2) give preference to residential developers, an increasing number of whom are absentee, even from out of state; (3) force local people to accept absentee interests; (4) limit—and perhaps even undermine—the scope of a local people's potential vision for sustainable community development within the context of their own landscape, especially for the desired future condition of their landscape; and (5) curtail or even eliminate the ability of local people to actively mourn for the continuing loss of their quality of life and their sense of place as outside choices are forced upon them, often by people who will not have to live with the consequences of their imposed actions.

The whole purpose of choice is for local people to guide the sustainable development of their own community within the mutually sustainable context of their landscape. After all, the local people and their children must reap the consequences of any decisions that are made. To limit their choices is to force someone else's consequences upon them, often at a great and increasingly negative long-term cost, first socially and then environmentally.

When preferential treatment is given to residential developers, including absentee developers, local people are at a serious disadvantage when it comes to planning for long-term community sustainability within the context of a finite landscape. While the focus of sustainable community development is long term, the interests of most residential developers are strictly short term, which usually counteracts long-term planning based on long-term environmental consequences. Further, it is exceedingly unlikely that absentee residential developers are going to have a vested interest in the long-term welfare of a community once they have made their money.

As noted in the above quote, it is the letter of the law that the residential developers want strictly enforced. But the letter of the law lacks moral consciousness, ignores the values of local residents, and discounts long-term planning for sustainable community development, all of which have consequences that are critical to the long-term social/environmental sustainability of a community. So, long after the residential developer is gone, the community is left to deal with the environmental errors caused by too much haste because the letter of the law was held to be inviolate and shielded from challenge.

Finally, as noted above, the expressed purpose of the law to limit public debate is to "get *emotions* [feelings, human values] out of land-use decisions," which effectively slaughters the quality of human relationships while enforcing the letter to the law for the benefit of residential developers. But emotions, the force behind relationships, are based on personal and collective values, which are the heart and soul of a community.

Protect and Enhance the Freedom of Public Debate

Debates over the use of a given piece of land, be it in Oregon, the United States, or elsewhere, are indeed fueled by personal values that are expressed as emotions. These debates, and the emotions they evoke, not only help the participants to integrate the proposed changes into their consciousness but also are a necessary and vital form of grieving over the imminent loss of a safe and known past and the invasion of an unknown and uncertain future.

Curtail public debate (which steals people's legal right to express their feelings through both statements and questions in exercising self-determination) and trust and emotional well-being wither like a dying leaf in a hot wind. To be healthy, people not only must be allowed to grieve but also must be given permission to grieve for their perceived losses, which is one of the functions of public debate. Only when people have moved through their grief is rational long-term planning for sustainable community possible.

Charles Darwin penned in 1872 that one "who remains passive when overwhelmed with grief loses [the] best chance of recovering elasticity of mind."[36] And Aldo Leopold in 1949 wrote: "For one species to mourn the death of another is a new thing under the sun."[37] Taken together, these two statements—both from people with scientific backgrounds—underscore that grief is both necessary and that it reaches beyond the loss of life in human terms.

Grief, although difficult at best, is vital to the emotional acceptance of a painful circumstance and to the reshaping of oneself in relationship to an outer world that reflects a new reality based on that painful circumstance. To this end, Elisabeth Kübler-Ross published a book titled *On Death and Dying*,[38] which simultaneously is a book on the *grief of life* and *living*. She described five stages that a terminally ill person commonly goes through when told of her or his impending death: denial and isolation, anger, bargaining, depression, and acceptance. Before relating these stages to our thought processes and how we change, let's examine the stages in terms of a dying person:

1. Denial, refusing to admit reality or trying to invalidate logic, is the first stage a terminally ill person goes through. Denial leads to a feeling of isolation, of being helpless and alone in the Universe. At some level, however, the person knows the truth but is not yet emotionally ready or able to accept it.
2. Anger, which is a violent outward projection of fear, can be called emotional panic. The person is emotionally out of control because she or he can no longer control heretofore controllable circumstances.
3. Bargaining is when a person attempts to bargain with God, however God is thought of, to change the circumstances, to find a way out of having to deal with what is.
4. Depression is a somewhat different type of issue because it comes in two stages. In the first stage, a person is in the immediate process of losing control of circumstances, such as a job and his or her identity with that job. The second stage is one in which a person is no longer concerned with past losses, such as

a job, but is taking impending losses into account, such as leaving behind friends, family, and perhaps pets.

5. Acceptance, the final stage, is creative and positive. With acceptance returns trust and a faith in the goodness and the justice of the outcome. Acceptance allows us to acknowledge our problem, which allows us to define it, which in turn allows us to transcend it. But first we must accept what is—the truth that sets us free.

Now let's apply these stages of grieving to the context of life and living. Although we are alive, we die daily to our ideas and belief systems, affected as they are by constant change. We are also faced with the death of beautiful things that we have long cherished, such as a small forested hill near our home, where we have spent many a happy hour over the years enjoying seasonal flowers, fresh breezes, and the silence of open space.

Suddenly, we are confronted with the prospect of a housing development (255 single-family houses, 230 apartments, and 38 townhouses—an actual case) on 103 acres of that hill, and our sense of impending loss and grief is acute. Now we begin to go through the stages of grieving that prepare the way for change over which we, the people, have less and less control because private interests are too often deemed more important than the public good of self-determination:

1. Denial of or resistance to change is how we isolate ourselves from one another because we see change as a condition to be avoided at almost any cost. We become defensive, fearful, and increasingly rigid in our thinking; we harden our attitudes and close our minds. If one becomes defensive about anything, starts to form a rebuttal before someone is finished speaking, filters what is said to hear only what one wants to hear, one is in denial.
2. Anger, the violent outward projection of uncontrollable fear, in the face of change renders one temporarily "insane": "I won't accept this!" One's anger, however, is not aimed at the person on whom it is projected; it is aimed at one's own inability to control the circumstances that seem so threatening or emotionally devastating.
3. Bargaining is looking for a way to alter the circumstances based on more "acceptable" conditions (to cut a better deal, if you

will), which is a function of public debate and the purpose of labor unions.

4. Depression is when one becomes resigned to one's inability to control or change the "system," whatever that is, to suit one's desires. One feels helpless and deliberately gives up trying to alter circumstances. One becomes a "victim" of "outside forces," and one's defense is to become cynical—to project forward one's distrust of human nature and motives, even when nothing bad has happened.

   A cynic is a critic who stresses faults and raises objections but assumes no responsibility. A cynic sees the situation as hopeless and is therefore a prophet of doom who espouses self-fulfilling prophecies of failure regardless of the effort invested in success.
5. Acceptance of the ultimate outcome allows one to transcend the purely emotional state and reach a point where emotion and logic integrate. In so doing, one can define the problem and in turn transcend it. But acceptance of the problem must come before transcendence is possible.

   Initially, however, we resist change because we are committed to protecting our existing values, representing as they do the safety of past knowledge in which there are no unwelcome surprises. We try to take our safe past and project it into an unknown future by skipping the present, which represents change and holds uncertainty, danger, and grief.

Ecologist Michael Soulé feels that: "As the number of exotics in most regions produces a cosmopolitanization of remnant wildlands, there will be an agonizing period of transition, especially for ecologists....There are moments when the destruction of a favorite place, of entire biotas and ecosystems, seems unbearable and the future looks bleak indeed."[39] But mourning for ecological losses, such as those Soulé expressed, or even for a small, private, spiritual sanctuary, say a wooded hill near your home, has a path that is neither simple nor predictable.

Those of us who have been trained to deal primarily through our intellect, which is but the first step in grasping the loss of someone or something we love, are too often cut off from our feelings and therefore try, as best we can, to minimize the pain. On the other hand, those who are in touch with their feelings and acutely aware of their pain are quickly accused by the monied interests, such as residential developers, of caring more for a wooded hill, wildflowers, or a butterfly than for people, which is implicit in the above-mentioned land-use law (House Bill 3065) pushed through the Oregon legislature in 1995.

It is no surprise, therefore, that the material world often makes grieving for the loss of our environment and our attachments to it a most difficult and uncertain process, where the need to defend personal values and the feelings they engender against cold materialism is all but a foregone conclusion. We have almost no social support for expressing grief. When one sits beside the bed in which a loved one is dying, it is both expected that and acceptable for people to cry about the unwelcome changes they are experiencing, including feelings that are dark and intense—rage at life's unfairness and guilt for doing too little. Our tears are a sign of grief work well done. But while honest conversations about grief may come quite naturally at the bedside of the dying, they are far more difficult and dangerous (less acceptable) at a conference table.[40]

Nevertheless, people have long used rituals to help themselves and one another mourn and recover from grief. Funerals and memorial services serve as a rite of passage between the initial shock of loss and the longer, more private and difficult phases of grieving. Not all customs of mourning are religious, however; take, for example, the group solidarity shown among mourners during open public debate concerning the use of land and/or the perceived degradation of the environment.[41]

Most of our customs of contemporary mourning are directed at the acute loss of the people and pets we love; these customs are important in the first weeks and months of the grieving process.[41] But environmental and social losses are intermittent, chronic, cumulative, and without obvious beginnings and endings.[40] It is therefore necessary to encourage, support, and develop (not curtail) safe customs of grieving for environmental and social losses, those which alter the context of our lives just a surely as the loss of a person or a pet.

Ancient customs as well as contemporary experts urge us to grieve both for its intrinsic benefits and because failure to grieve can have incredibly far-reaching consequences. Failure to grieve for the environment, for example, can become chronic grief (often the source of cynicism) from which recovery seems never to come. Alternatively, we can inhibit grief, which then becomes distorted and may erupt into acts of frustration and violence.

Although Charles Darwin concluded in 1872 that grieving serves us well in the long run,[36] Colin Murray Parkes expanded this notion. Our willingness to look at grief and grieving, he wrote, instead of turning away, is the key to successful grief work. We may choose to deal with our fear by turning from its source, but each time we do, we only add to our fear, perpetuate our problems, and miss an opportunity to prepare ourselves for the inevitable changes in a changing world.[42]

"Fortitude," says Michael Soulé, is what we need "when the temptation to turn and walk away is almost overpowering." Fortitude is the heart of not only grieving but also local government. The fruits of courage become self-reinforcing feedback loops.

Protect and Enhance Information Feedback Loops

Information is the fundamental ingredient in the process of creating new structures within a community, where complex new forms of structure come into being simply by information feeding back on itself. Sustainable community development thus creates a mechanism for information to feed through the political system and direct change toward a dynamic equilibrium between the community and its environment. In the case of adjacent communities, a collective mechanism of information feedback is also essential if sharing the same landscape and its common products, such as water, is going to be just and sustainable.

People's values, belief in process, and the empowerment to act and collectively resolve problems are the first component of sustainable community development. Education that allows people to learn of their connection to social/environmental problems, both local and global, is the second part of the mechanism. Teaching participants how to plan strategically is the third part, and a sound working knowledge and practice of the democratic system of government within which to fit the first three parts is the final, all-encompassing piece. It is all-encompassing because its processes of public representation funnel people's knowledge, feelings, requirements, desires, and concerns into an informational feedback loop that directs change. In this manner, sustainable community development can have a significant effect in directing societal change toward environmental sustainability.

Sustainable community development can instill a sense of purpose and belonging that is defined and maintained by a local community within its environmental context. It does this by integrating all aspects of society in working toward a dynamic balance of sustainable outcomes.

Such balance can only be maintained if information is fed back into the system in a way that fosters new questions and new practices appropriate for changing circumstances while simultaneously discarding only inappropriate old questions and old ways. Sustainable community development is based on information feedback, not social insanity, which is trying the same old thing over and over while each time expecting a different and more desirable outcome.

Wise and sustainable development requires our total conscious presence in the present. It also requires that decisions be based on true human/environmental indicators (such as feelings that are balanced with logic and intuition, between questions based on social value and questions based on scientific understanding, and between social and environmental necessities; self-worth; responsibility in the present for the future; community [including economic]/ecological adaptability; and so on).

This may sound easy, but it is extremely demanding in terms of concentration and energy. The most difficult part of developing community/environmental sustainability is that we will all have to let go of some of our old, cherished beliefs and desires, such as the long-held simplistic notion that economic development is the sum total of community development.

And we will have to face new challenges as we alter the ecosystem within which we live and participate, thereby making it ever-more fragile and our relationship to it ever-more labor intensive and energy demanding. Consider, for example, that in remote times, the nomadic peoples, driven from their homes by scarcity of game and/or impoverished soil, migrated sometimes great distances in search of food and water. In so wandering, the peoples of old escaped many of the diseases that afflict modern society.

But when people began settling into more permanent communities, their health problems increased immediately. As wanderers, they had left their refuse behind and moved into healthier environments, but as city builders, they piled their refuse at the outskirts of the villages and towns, and there was born a great part of the sickness that still plagues society.

Although the problems of sanitation have been largely solved in the rural communities and large cities of the wealthy industrialized nations, such is not the case in much of the world. And even a wealthy industrialized nation like the United States still has its stubborn problems. We say stubborn because they are ignored as much as possible by that part of the populace interested more in personal financial gain than in human/environmental welfare, present or future.

Sanitation in the days of nomadic peoples required one set of information feedback loops. Addressing the problem of sanitation with the onset of agriculture and the rise of permanent communities required a new set of information feedback loops. Today, the complexity of human communities within the context of severely altered landscapes influenced from afar by a still larger society requires still another, more comprehensive systems-oriented set of information feedback loops. These new feedback loops must be protected and enhanced by local government if local communities are to learn from their failures and successes so they can more easily adapt to ever-changing circumstances.

Increase Local Adaptability

Societies around the globe are in the throes of change, some because of diminishing resources and others because of social upheaval, but all are losing control over their destinies. Regardless of the cause, sustainable local community development provides vision, planning, and direction in times of crisis as well as in times of peace.

The focus of the local government under the auspices of sustainable community development must be on balancing the ability of a community to meet its own needs while maintaining relative economic stability as outside markets fluctuate. This must be done while simultaneously protecting the ability of future generations to meet their needs in the same area.

Sustainable community development works to maintain a dynamic equilibrium through consciously directed, systemic, self-reinforcing feedback loops. It offers a process that can mobilize citizens to direct information toward long-term community sustainability, which in some measure equates with the economic stability of a local community.

Sustainable community development increases the adaptability of a community by creating and maintaining a diversified social and economic base with local shared ownership and access to basic human services. Community adaptability, and therefore stability, is based on the ability of the community to meet the majority of its own needs within itself instead of being dependent on outside resources. This means, however, that the adaptability of a community also encompasses the ecological integrity of its surrounding landscape. How might this be done?

Agencies with jurisdiction over public lands could write performance-based contracts that set overall sustainable ecological outcomes for a forest, grazing allotment, or water catchment while allowing flexibility in finding ways to achieve the objectives. With careful attention to incorporating components of accountability and requirements for broad community participation, trusteeship contracts could allow community partnerships to help "manage" natural resources for future generations while helping to meet local, regional, state, and national needs from our public lands.

Such trusteeship contracts will require new levels of trust. Trust among government agencies, trust given by such agencies to community partnerships, and trust between environmental and industrial interests are all requisite to achieving trusteeship contracts. The initial effort to create a program of trusteeship contracting must begin small, with little successes, and move to more and greater devolution of authority as a community and its partnerships mature and accomplish or exceed the agreed-upon measures of performance. The level of trust among all parties will, of course, grow with each success.

As government agencies perform their responsibilities of caring for public lands, whether federal, state, county, or city, they must plan any and all economic activities that nurture the local economy to fit within the biologically sustainable capabilities of the land over time. For example, a mill worker who has been laid off permanently because the local mill closed must be given preference in any opportunity to help restore the biological health of the forest from which the mill's logs once came. In turn, revenues from the forest must remain, for as long as possible, within the forest and local area as rollover investments and reinvestments in the land and people, which are inseparable partners in reciprocal feedback loops of the ecosystem.

We need new partnerships and we need trusteeship contracts between government at all levels and resource-based communities. Such partnerships and contracts are necessary to help resource-based communities as they struggle with diversification for a sustainable future. These partnerships and contracts will, however, require achieving specific ecological and economic results within measures of agreed-upon accountability from within the community partnerships themselves and from all relevant levels of government.

As public values shift from using natural resources strictly for extraction to recognizing the linkages between the long-term health of ecosystems and the quality of life that healthy ecosystems allow, people in local communities must be given the opportunity to earn livelihoods from new industries, such as ecosystem restoration, recreation, or specialty renewable products. The landscape is a community's spiritual as well as capital asset and must be considered the foundation upon which a community's cultural identity and economic sustainability rest.

As such, it is the community's moral responsibility through citizen participation to safeguard the ecological sustainability of its surrounding landscape as its legacy to those who must follow us in our ever-shrinking world with its burgeoning human population. This responsibility, however, will require a shared vision of a sustainable future.

10

A SHARED VISION: THE GATEWAY TO A COMMUNITY'S FUTURE

Eyesight is nothing without vision, yet the meaning and purpose of a collective vision as a shared experience is something few people seem to understand, and because they fail to understand it, many people simply take potluck with their future. When asked about their future, however, people often say, "I have a vision of what I want." But when pressed to explain their vision, it becomes clear that they have not the foggiest notion that a vision is a strong organizing context in the present for the future. A vision consists of what people ideally want, not of what they fear.

Be that as it may, why does a vision work? A carving from a church in Sussex, England, suggests an answer: "A vision without a task is but a dream. A task without a vision is drudgery. A vision and a task are the hope of the world." A shared vision of a sustainable future toward which a community can build creates confidence, consensus, and energy in equal parts.[43] At a deeper level, it engages our imagination and helps to ferret out which questions need to be asked, how to word them, and when to ask them.

By engaging our imagination and our sense of possibility of the ideal through countless small-scale initiatives, such as shared community visions, people who are concerned with the health of their environment and social justice can create an opportunity to confirm a more positive sustainable future. Imagination, as Albert Einstein said, is more important than knowledge and is the most powerful tool for social change.

We must change our values and habits if we are to be ecologically sustainable and socially inclusive in our ways of living. A shared vision of the future starts the process of change by enabling people to think differently about their lives and in so doing commences to change them.

A resident community, through its vision, accepts responsibility for its own survival, and outsiders must fit themselves into that vision if the community is to be sustainable. A vision can thus be likened to a human body in that the strong organizing context of the body's division of labor keeps the various cells functioning within acceptable bounds.

If, for example, cancer cells are removed from a cancerous animal, where they are defying the weakened organizational context of the body, and are placed in a healthy animal, they return to a normal functional pattern because they come under the influence of a strong organizing context. If, however, cells stay too long without the guidance of a strong organizing context, they reach a point where they can no longer be guided at all, and they become "rogue" cells or "cancer" cells.

If the context is disorganized, confused, or murky, then it is not surprising that the cells, whether of the human body, the body of a human community, or the body of humanity as a whole, tend to go out of control. The longer disorganization exists, the more difficult it is to reverse its effects. Hence a community's need for the strong organizing context of a vision, the examination of which is the purpose of this chapter.

An ancient custom of the First Americans was to call a council fire when decisions affecting the whole tribe or nation needed to be made. To sit in council as a representative of the people was an honor that had to be earned through many years of truthfulness, bravery, compassion, sharing, listening, justice, being a discreet counselor, and so on. These qualities were necessary because a council fire by its very nature was a time to examine every point of view and explore every possibility of a situation that would in some way affect the whole of the people's destiny. In today's terminology, this might be called a "visioning process."

When someone called a council, that person had to have the courage to accept the council's decision with grace, because when the good of the whole is placed before the good of the few, all are assured a measure of abundance. The timeless teaching of the council fire is that until *all* of the people are doing well, *none* of the people are doing well.

Borrowing from the days of old, the modern visioning process is designed around the same timeless teaching of the ancient council fire, namely that until *all* of the people are doing well, *none* of the people are doing well. Thus, for the sake of sustainability, present and future, the good of the whole must be placed before the good of the few.

The council fire worked well for the First Americans because they knew who they were culturally, and they had a sense of place within their environment. Today, however, societies around the globe are in transition, which robs many people of their original sense of place and substitutes some vague idea of location. With so many people feeling adrift in their lives, the whole concept of a vision, much less one that includes future generations, seems inconsequential at best if not an exercise in abject futility.

This transition is largely the result of massive shifts in human populations over the last three centuries. These shifts have altered the composition of peoples and their cultural structures throughout the world. All of this activity results in growing interconnectedness, interactivity, interdependence, and cultural uncertainty as some political lines change physically and others blur culturally.

Cultural uncertainty is particularly true for those people caught between two cultures, such as the warring religious factions around the world, where millions of refugees not only have their sense of culture disrupted but also have their sense of place transformed into that of an alien location in which their lives hang in limbo. A large number of these people are immigrating to the United States, where many are trying to reconstruct the foundations of their own cultures while fitting compatibly into a new culture and thus a new sense of community.

These shifts in population are forcing even some of the most parochial communities to see themselves from new and different points of view. Others are being forced to look at themselves anew because today's social/environmental conditions—driven by technology—are changing so fast that many of our known and comfortable self-views are obsolete. This being the case, there are some preliminary subjects, beginning with our own perception of our community and world, that need examination prior to dealing with the notion of a vision itself.

THROUGH THE EYES OF AN INSECT

When Margaret Shannon, a professor of natural resource policy and sociology at the State University of New York, said, "The world does not define itself for us; rather we choose to see some parts of the world and not others," she opened the door to a whole new way to think about culture: that of our individual and collective perceptions. Her statement puts us on notice that we do not *see* clearly our own culture, but rather we have some perception of it—which in itself creates our culture, because my perception is more or less different from yours, sometimes vastly so.

The amount of chaos and conflict in a community is therefore a direct measure of how different people's perceptions really are and how committed they are to defending their individual points of view, regardless of their narrowness in scope. The purpose of a vision is to render the present chaos into the greatest possible harmony for the collective benefit of the community as a whole through time, recognizing that each person's perception is part of the community's living culture.

Living culture is thus embodied in the people themselves, and it is there one must search for an understanding of a people as a whole. In this sense, each person is both the creator and the keeper of a unique piece of the cultural tapestry, an understanding of which one can glean only by seeing it simultaneously from many points of view—much as an insect sees.

Our perceptions can be thought of in a manner similar to that of an insect's compound eyes because it is through perception that we "see" one another and everything else. An insect's compound eyes are formed from a group of separate visual elements, each of which corresponds to a single facet of the eye's outer surface, which may vary from a few hundred to a few thousand, depending on the kind of insect. Each facet has in turn what amounts to a single nerve fiber that sends optical messages to the brain. Seeing with an insect's compound eyes would be like seeing with many different eyes—or perceptions—at once.

Each perception of a component of one's community is like a facet in the compound eye of an insect, with its independent nerve fiber connecting it to the local community and hence expanding outward to the regional, national, and global society (the various levels of our increasingly collective and abstract brain). Thus, each perception, being composed of elements of many things, including an individual's personal and cultural foundation, has its unique construct. This of course establishes the limits of an individual's understanding.

A person who tends to be positive or optimistic, for example, sees a glass of water as half full, while a person who tends to be negative or pessimistic sees the same glass of water as half empty. Regardless of the way it is perceived, the level of water is the same—illustrating that we see what we choose to see, which has everything to do with perception but may have little to do with reality.

The important implication is that the freer we are as individuals to change our perceptions without social resistance in the form of ridicule or shame, the freer is a community (the collective of individual perceptions) to adapt to change in a healthy, evolutionary way. On the other hand, the more people are ridiculed or shamed into accepting the politically correct ideas of others, the more prone a community is to the cracking of its moral foundation and to the crumbling of its social infrastructure, because social change cannot be held in abeyance for long, which poses questions to which we must respond.

QUESTIONS WE NEED TO ASK

Before the people of a community are ready to craft a shared vision of their future, they must ask and answer two questions: (1) Who are we today as a culture? and (2) What legacy do we want to leave our children?

Who Are We as a Culture?

Who are we culturally—now, today? This is a difficult but necessary question for people to deal with because a vision is the palpable nexus between a fading memory of the past and the anticipation of an uncertain future. The people of a community must therefore decide, based on how they define their present cultural identity, what kind of vision to create. A people's self-held concept (individual, cultural, and universal values) is critical to their cultural future because their personal and cultural self-image will determine what their community will become socially, which in turn will determine what their children will become socially.

Thomas Jefferson gave good counsel on values: "In matters of principle, stand like a rock. In matters of taste, swim with the current." How do we identify those principles and/or values on which we stand firm? Well, we can ask ourselves: What are the fundamental principles that I believe in to the point of no compromise? What values are central to my being?

Categories of Value

The Ch'an masters who carried Zen to Japan brought Confucian ethics with them. In discussing these fundamental values as a guide to personal behavior, Confucius said: "If a man will carefully cultivate these in his conduct, he may still err a little, but he won't be far from the standard of truth."[44,45] When we as individuals clearly understand and can explicitly articulate our personal values, then we can live in keeping with them.

Let's consider three categories of values: universal, cultural, and individual.[46] Universal (or archetypal) values reveal to us the human condition and inform us of our place therein. Through universal values, we connect our individual experiences with the rest of humanity (the collective unconscious) and the cosmos. Here, the barriers of time and place, of language and culture disappear in the ever-changing dance of life. Universal values must be experienced; they cannot be comprehended. Can you, for example, know a sunset? Fathom a drop of water? Translate a smile? Define love?

Universal values are the timeless constants brought to different cultures at various times throughout history. "Even as the hands of a clock are powered from the center, which remains ever still, so the universal values remain ever at the center of human life, no matter where the hands of time are pointing—past, present, or future."[46] These are the truths of the human condition toward which people aspire (such as joy, unity, love, and peace); of these the sages have spoken in many tongues.

Cultural (or ethnic) values are those of the day and are socially agreed upon. They are established to create and maintain social order in a particular time and place and can be highly volatile. Cultural values concern ethics and human notions of right and wrong, good or evil, in terms of customs and manners.

In culture we see reflected the ideas and behaviors that a society rewards or punishes according to their perceived alignment with its values. Hence, cultural values are for an individual a mixed bag, especially in a highly complex society that has lost its sense of family, community, and mythology, like that of the United States, where there is much that may resonate with an individual and much that may not.

Every culture is a person in a sense, and like people, there is the potential for creative interaction and/or conflict when cultures meet. Although we are all too familiar with cultural conflicts and the destruction they have wrought, it is well to remember that a meeting of cultures also triggers tremendous explosions of creativity in such things as language, ethics, education, law, philosophy, and government.

Individual (or personal) values are constituted by the private meanings we bestow on those concepts and experiences (such as marriage vows or spiritual teachings) that are important to us personally. These meanings are in large part a result of how we are raised by our families of origin and what of our parents' values we take with us in the form of personal temperament. These meanings may change, however, depending on our experiences in life and how much we are willing to grow psychologically and spiritually as a result of our experiences. As such, individual values are reflected in such things as personal goals, humor, relationships, and commitments.

Thus, how well a people's core values are encompassed in a vision depends first on how well the people understand themselves individually and as a culture, which means how well they understand their core values, and second on how well that understanding is reflected on paper, where there can be no question about what has been stated and how. Let's consider the First Canadians.

The First Canadians have departed from their old culture because they have—against their will—been forced to adopt European–Canadian ways, which means they have given up or lost ancestral ways. Yet they have not—by choice—totally adopted white culture and want to retain some degree of their ancestral culture. Thus, the three questions they must ask and answer are: Which of our ancestral ways still have sufficient cultural value for us to keep them? Which of the white ways do we want to adopt? How do we put the chosen elements of both cultures together in such a way that we can today define who we are culturally?

For example, in 1993, Chris was asked to review an ecological brief for a First Nation in Canada whose reservation is located between the sea and land immediately downslope from that which a timber company wanted to cut. The problem lay in the fact that the timber company could only reach the timber it wanted to cut by obtaining an easement through the reservation, which gave the First Nation some control over the timber company. The First Nation wanted this control to have an active voice in how the timber company would log the upper-slope forest, because the outcome would for many years affect the reservation, which is immediately below the area to be cut.

The First Nation was in this case the strong organizing context (by virtue of the company's required easement through the First Nation's land) that would control the behavior of the timber company as it logged the upper-slope forest. If, however, the timber company had not been required to pass through the First Nation's land, it could, through self-serving logging practices, easily have become the uncontrollable cancer that would have destroyed the cultural values of the First Nation's land for many generations.

Before meeting with the timber company, the First Nation's chief asked for some counsel. Chris's reply was as follows:

> Before I discuss the ecological brief I've been asked to review, there are three points that must be taken into account if what I say is to have any value to the First Nation. What I'm about to say may be difficult to hear, but I say it with the utmost respect.
>
> **Point 1:** Who are you, the First Nation, in a cultural sense? You are not your old culture because you have—against your will—been forced to adopt some white ways, which means you have given up or lost ancestral ways. You are not—by choice—white, so you may wish to retain some of your ancestral ways. The questions you must ask and answer are: What of our ancestral ways still have sufficient value that we want to keep them? What of the white ways do we want to or are we willing to adopt? How do we put the chosen elements of both cultures together in such a way that we can today define who we are as a culture?
>
> **Point 2:** What do you want your children to have as a legacy from your decisions and your negotiations with the timber company? Whatever you decide is what you are committing your children, their children, and their children's children to pay for as the effects of your decisions unto the seventh generation and beyond. This, of course, is solely your choice and that is as it should be. I make no judgments. But whatever you choose will partly answer Point 3.
>
> **Point 3:** What do you want your reservation to look like and act like during and after logging by the timber company? How you define yourselves culturally, what choices you make for your children, and the conscious decisions you make about the condition of your land will determine what you end up with. In all of these things, the choice is yours. The consequences belong to both you and your children.

What about you, the reader—who are you today? We each change personally as we grow in years and experience. So do our respective communities. Each community that wishes to create a vision for a sustainable future must therefore ask of itself: Who are we today in a cultural sense? Then, based on how a community sees itself, each community must ask: Who do we want to be or to become in the future? These are important questions, which must be clearly answered on paper for all to see, because how they are answered will determine the overall direction of a community's vision and thus the legacy inherited by its children.

Ferreting Out Community Values

To ask a relevant question about where you are going, you must know not only where you want to go but also where you are, which means taking stock of who you are. Whereas a shared vision is a statement of where you want your community to go, assessing your community, including the reciprocity of its relationship to the immediate landscape, as it is today allows you to determine your starting point for the journey.

One way of assessing a community is by entering into its routines.[47] This means having selected people attend school events, such as football games and meetings of the Parent–Teacher Association; visiting people in their kitchens and living rooms; and going into cafes, gas stations, laundromats, and other places where people gather, such as taverns and churches. The purpose of these visits is to interact with residents to determine such things as what they do for work and what their work routines are, their personal interests, recreational patterns, what support services are important to them, and how they feel about changes within the community and between the community and its landscape.

To really understand how a community sees itself, one must ask people not only what they like about their community and its landscape and why but also what they do not like and why. One must ask people what they most want to change about their community and its landscape. Questions also help one find out which informal networks people use both to communicate with one another and to solve problems, as well as whom they trust and rely on as communicators and caretakers.

The purpose of asking such questions is to make the informal system of community clearly visible in such a way that by understanding the range of issues people are concerned about and how they see themselves in relationship to those issues, one can help the community recognize and express its current cultural identity. This kind of information is called "ethnography" in anthropology, or "the story of the people."

The story of the people as a baseline description of how the people identify themselves culturally is a sound preparatory step toward crafting a shared vision. A sustained process of interaction within a community at the informal level has two important effects.

One, it fosters empowerment of the people themselves and as a community because personal and social reflection not only determines the intelligence and possible consequences of any given action but also leads citizens to see what the next step might be and to take it. It is thus important, as French philosopher Henri Bergson observed, to "Think like a...[person] of action, and act like a...[person] of thought."

Two, it can prompt social institutions to become more responsive because people within agencies gain insight into the concerns of citizens and thus into a community's cultural identity by participating in the ongoing "story of the people." Such participation both gives agency people good and relevant information that makes sense to the citizens and allows them to understand why citizens say what they do. This notion is reminiscent of a statement made by Mahatma Gandhi to an audience of India's bureaucrats and social elite: "Until we stand in the hot sun with the millions that toil each day in the fields, we will not speak for them."

Such small-scale change, done with and by people rather than for or to them, when multiplied over a whole community, becomes a clear signpost toward a community's vision of social/environmental sustainability and hence the legacy people leave their children.

What Legacy Do We Want to Leave Our Children?

Once a group of people, whether a community, such as an indigenous peoples or your own hometown, has defined itself culturally (present and future), it can decide what legacy it wants to leave its children. This must be done consciously, however, because the consequences of whatever decisions the group makes under its new cultural identity are what the group is committing its children, their children, and their children's children to pay.

The rest of Chris's reply to the First Nation in Canada applies here:

> Now to my comments: This is a difficult task at best. As with any definition, it is a human invention and has no meaning to Nature. Therefore, you must tell the timber company, clearly

and concisely, what the terms in this ecological brief mean to you and how you interpret them with respect to the company's actions that will affect your reservation.

1. Every ecosystem functions fully within the limits (constraints) imposed on it by Nature and/or humans. Therefore, it is the type, scale, and duration of the alterations to the system—the imposed limits—that you need to be concerned with.

If your reservation looks the way you want it to and functions the way you want it to, then the question becomes: How must we and the timber company behave to keep it looking and functioning the way it is? If, on the other hand, your reservation does not look the way you want it to and does not function the way you want it to, then the question becomes: How must we and the timber company behave to make it look and function the way we want it to?

But regardless of your decisions or the company's actions, your reservation will always function to its greatest capacity under the circumstances (constraints) Nature, you, and the company impose on it. The point is that your decisions and the company's actions, excluding what Nature may do, will determine how your reservation both looks and functions. This reflects the importance of the preceding Point 3 and what you decide.

2. If you want the landscape of your reservation to look and function in a certain way, then how must the timber company's landscape look and function to help make your reservation be what you want it to be? Keep in mind that the landscape of your reservation and the company's timber holdings are both made up of the collective performance of individual stands of trees or "habitat patches." Therefore, how the stands look and function will determine how the collective landscape looks and functions.

3. Remember that any undesirable ecological effects are also undesirable economic effects over time. Your interest in your reservation will be there for many, many years, generations perhaps, but the company's interest in the forest may well disappear just as soon as the trees are cut. So, the company's short-term economic decision may be good for them immediately but may at the same time be a bad long-term ecological and thus a bad long-term economic decision for you.

4. To maintain ecological functions means that you must maintain the characteristics of the ecosystem in such a way that its processes are sustainable. The characteristics you must be concerned about are: (1) composition, (2) structure, (3) function, and (4) Nature's disturbance regimes, which periodically alter the ecosystem's composition, structure, and function.

The composition or kinds of plants and their age classes within a plant community create a certain structure that is characteristic of the plant community. It is the structure of the plant community that in turn creates and maintains certain functions. In addition, it is the composition, structure, and function of a plant community that determines which animals can live there and how many. If you change the composition, you change the structure, you change the function, you affect the animals. People and Nature are continually changing a community's structure by altering its composition, which in turn affects how it functions.

For example, the timber company wants to change the forest's structure by cutting the trees, which in turn will change the plant community's composition, which in turn will change how the community functions, which in turn will change the kinds and numbers of animals that can live there. These are the key elements with which you must be concerned, because an effect on one area can—and usually does—affect the entire landscape.

Composition, structure, and function go together to create and maintain ecological processes both in time and across space, and it is the health of the processes that in the end creates the forest. Your forest is a living organism, not just a collection of trees—as the timber industry usually thinks of it.

5. Scale is an often-forgotten component of healthy forests and landscapes. The treatment of every stand of timber is critically important to the health of the whole landscape, which is a collection of the interrelated stands. (A stand of trees is a delineated group of standing trees.)

Thus, when you deal only with a stand, you are ignoring the relationship of that particular stand to other stands, to the rest of the drainage, and to the landscape. It's like a jigsaw puzzle where each piece is a stand. The relationship of certain pieces (stands) makes a picture (drainage). The relationship of the pictures (drainages) makes a whole puzzle (landscape). Thus, the relationships of all the stands within a particular area make a drainage, and the relationships of all the drainages within a particular area make the landscape.

If one piece is left out of the puzzle, it is not complete. If one critical piece is missing, it may be very difficult to figure out what the picture is. So each piece (stand) is critically important in its relationship to the completion of the whole puzzle (landscape). Therefore, the way each stand is defined and treated by the timber company is critically important to how the landscape, encompassing both the company's land and your reservation, looks and functions over time.

6. Degrading an ecosystem is a human concept based on human values and has nothing to do with Nature. Nature places no extrinsic value on anything. Everything just is, and in its being it is perfect (intrinsic value). Therefore, when considering intrinsic value, if something in Nature changes, it simply changes—no value is either added or subtracted. But superimposing the extrinsic value of human desires on Nature's intrinsic value creates a different proposition. Thus, whether or not your reservation becomes degraded depends on what you want it to be like, what value or values you have placed on its being in a certain condition, to produce certain things for you. If your desired condition is negatively affected by the company's actions, then your reservation becomes degraded. If your desired condition is positively affected by the company's actions, then your reservation is improved. Remember, your own actions can also degrade or improve your reservation.

7. It is important that you know—as clearly as possible—what the definitions in this brief really mean to you and your choices for your children and your reservation. Only when you fully understand what these definitions mean to you can you negotiate successfully with the timber company.

If you, the reader, substitute the name of your own community wherever "reservation" occurs in the above brief, and if you substitute "land-use zoning" or "land-use planning" wherever "timber company" occurs, you have an outline to follow for your own community, perhaps with a few modifications to fit specific local conditions.

To negotiate effectively with the timber company, however, the First Nation must have a vision, goals, and objectives for its reservation. For your community to negotiate effectively with the future, it too must have a clear vision, goals, and objectives.

VISION, GOALS, AND OBJECTIVES

Having defined who they are culturally and having determined what legacy they want to leave their children, the people of a community are now ready to craft a vision of what they want because only now do they really know. Although the word "vision" is variously construed, it is used here as a strong organizing context in the form of a shared view of the future, which is based on its three separate but overlapping aspects: world view, perception, and imagination.[46]

World View

Our world view is our way of seeing how the world works, our overall perspective from which we interpret the world and our place in it. But it can also be seen as a metaphysical window to the world, which cannot be accounted for on the basis of empirical evidence any more than it can be proved or disproved by argument of fact. "Metaphysical" simply means "beyond" (*meta*) the "physical" (*physic*), of which Albert Einstein said: "The more I study physics, the more I am drawn to metaphysics."

There are in the most general terms two world views: the sacred and the commodity. One need not be religious in the conventional sense to hold a sacred view of life, because a sacred view focuses on the intrinsic value of all life. As such, it gives birth to feelings of duty, protection, and love while emphasizing the values of joy, beauty, and caring, which in turn erects *internal* constraints to destructive human behavior against Nature. "Sacred" comes from the Latin *sacer,* which has the same root as *sanus,* "sane." A sacred view of life is therefore a sane view, which corresponds to the Sanskrit *sat, cit, ananda*—or being, consciousness, and bliss.

A commodity view of life is one interested in domination, control, and profit and which seeks to "gain the world" by subjugating it to the will of the industrial mentality. At the core of the commodity world view are several economic seeds, such as self-interest, the economy versus ecology (EVE) dilemma, growth/no-growth tug-of-war, Rational Economic Man, and others, which have already been discussed in some detail. It is necessary with respect to a commodity world view to protect the health of the environment in the present for the present and the future through *external* constraints placed on destructive human behavior.

If, therefore, we are going to change life, to improve it in any appreciable way, we must begin with attitudes, not facts. An outer change always begins with an inner shift in attitude, which Albert Einstein called "a new level of thinking."

"The world we have made," Einstein said, "as a result of...[the] level of thinking we have done thus far creates problems we cannot solve at the same level at which we created them." If, therefore, we shift our view of life from the profane to the sacred, we create a new set of facts and conditions as well as a new perception of the old facts and conditions because it represents a new level of thinking and thus of perceiving.

Perception

Our perception is the vision with which we see the world and interpret what we see in that we create our own world by both our attitudes toward it and our perceptions of it. Perception comes from perceive, which is from the Latin *percipere,* to seize wholly. "It is one of the great marvels of consciousness," writes author Laurence Boldt, "that whatever situation we clearly perceive, we improve."

To see wholly is to see a better way; therefore, to perceive a problem clearly is to begin formulating its solution. To solve a problem or resolve a conflict, we need the wisdom to keep searching and the confidence of love to hold what we find up to the light of understanding. It is when we doubt our capacity to love and to create, which we then replace with fear and isolation, that we begin to distort our perceptions of the world. When our view of the world is based on the love and confidence of clear perception, the world becomes a better place. It cannot be otherwise.

Nevertheless, we are usually moved from avoidance and confusion to attention and clarity only when we perceive the necessity to do so. "Necessity," wrote Plato, "is the mother of invention." Necessity, in this sense, is simply the perception that a current situation has become intolerable and that something must be done about it, which means the perceiver is the one who must act.

Once necessity is acknowledged and *accepted,* we begin searching for a solution to our problem by examining our old perceptions, which forces us out of our current prejudices and conceptual limitations in such a way that we can sift through those old ideas and concepts that we have in the past overlooked and/or discarded. Fortunately, necessity has an intense urgency about it, without which we too easily and too often give up searching for a better way of being or doing.

"Of course," Laurence Boldt says, "necessity, like beauty, is in the eye of the beholder." No matter how deplorable a situation seems to an observer, the individual in the situation finds no reason to improve it until such improvement becomes for him or her a personal necessity. Until our discontent is moved to necessity and we demand a better way, we will accept that which is of lesser quality. Because we fear and thus hate the things that seem to trap us, we find no way out until we supplant our fear with love and its counterpart, confidence.

We are stuck with our perceived problems until we have the determination, strength, and confidence to view them with love; this applies to problems on the scale of the Earth and our respective communities, as well as to our own personal troubles. The gift of the discoverer and innovator is the ability and tenacity to keep focused on what most of us avoid, the often lonely search for a better way, about which historian Daniel J. Boorstin writes: "The obstacles to discovery—the illusions of knowledge—are also part of our story. Only against the forgotten backdrop of the received common sense and myths of their time can we begin to sense the courage, the rashness, the heroic and imaginative thrusts of the great discoverers. They had to battle against the current facts and dogmas of the learned."[17]

By the very questions he or she asks, the discoverer and innovator elevates finding a new way to a necessity to which she or he is committed in a personal quest. Yet time and again we accept limitations because, not knowing what questions to ask or how to ask them, we cannot see the alternatives in front of our noses.

"Remember," says Laurence Boldt, "people were once told that bleeding was the best cure for disease and that slavery was an economic necessity." And in our present generation, we are told that we must prostitute our principles to get along and that many, if not most, of the problems in the world today are all but insurmountable. These notions call forward our individual and collective choices, namely, yield to the comfortable blindness of ignorance or summon our courage and make resolute our determination to search until we find a better way. It has been wisely said that anything will reveal its secrets if you love it enough.

Imagination

Imagination, or seeing that which can be, is the third aspect of vision. Even as we open our physical eyes and see the world as it is, with all its problems and opportunities, so we can open the eye of our mind and see the possibilities of as yet unseen realities. To open our physical eyes fully, we must learn to trust that we can accept what is through the eyes of love. To open the eye of our mind, we must learn to trust that what we see in our imagination we can bring forth in the physical world.

Whereas perception involves seeing that which already exists in the outer world, imagination involves seeing in one's inner world the possibilities that can be made manifest in one's outer world. Albert Einstein penned it nicely: "Your imagination is your preview of life's coming attractions." Consider, therefore, that everything humanity has ever created (or ever will create), both tangible and intangible, began as a single idea in the privacy of someone's mind, be it this book, a religious order, or going to the moon. Our imagination is the source of our power to create and the driving force behind our choices—the prerequisites of a shared vision toward which to build.

Understanding a Vision, Goals, and Objectives

Defining a vision and committing it to paper goes against our training because it must be stated as a positive in the positive, something we are not used to doing. Stating a positive in the positive means stating what we mean directly. For example, a local community has an urban growth boundary that it wants to keep within certain limits, which can be stated in one of two ways: (1) we want our urban growth boundary to remain within a half a mile from where it is now situated (a positive stated as a positive) or (2) we don't want our urban growth boundary to look like that of our neighbor (a negative that one is attempting to state as a positive).

Further, to save our planet and human society as we know it, we must be willing to risk changing our thinking in order to have a wider perception of the world and its possibilities, to validate one another's points of view or frames of reference. The world can be perceived with greater clarity when it is observed simultaneously from many points of view. Such conception requires open-mindedness in a collaborative process of intellectual and emotional exploration of that which is and that which might be, the result of which is a shared vision of a possible future.

There are two sayings that are pertinent here: If you don't know where you're going, any path will take you there, and if you stand for everything, you soon find that you stand for nothing. Thus, without a vision, we take "potluck" in terms of where we will end up, which was Alice's dilemma when she met the Cheshire-Cat in Lewis Carroll's story of *Alice's Adventures in Wonderland*.[48] Alice asked the Cheshire-Cat:

> "Would you tell me, please, which way I ought to go from here?"
>
> "That depends a good deal on where you want to get to," said the Cat.
>
> "I don't much care where—" said Alice.
>
> "Then it doesn't matter which way you go," said the Cat.
>
> "—so long as I get somewhere," Alice added as an explanation.
>
> "Oh, you're sure to do that," said the Cat, "if you only walk long enough."

The movie *Spartacus,* which depicts the true story of a Roman slave who as a boy of thirteen had been sold into slavery, is an excellent illustration of the power of a collective vision. Bought as a young man, Spartacus was taken to a highly organized school, where he was forced to learn fighting and become a gladiator. There was, however, a revolt early in his career and he, along with his fellow gladiators, escaped.

For a time, they ran roughshod over the countryside, disorganized and out of control, like the cancer cells discussed at the beginning of this chapter. They robbed, raped, and murdered the Roman gentry and encouraged their slaves to join the growing mob. But Spartacus was uncomfortable with the barbaric mob because he recognized that it had simply become what it was against; it had become like the Romans. He therefore organized the slaves into an army that would fight its way across Italy to the sea and escape.

Thus, in 71 B.C., Spartacus led his army in an uprising. Now a highly organized fighting machine that opposed Roman rule, Spartacus's army had become a dangerous, out-of-control cancer (by Roman standards) that threatened the Roman sense of superiority, because it was, after all, just an army of slaves. Although the slaves twice defeated the Roman legions, they were finally conquered by General Marcus Licinius Crassus after a long siege and battle in which they were surrounded by and had to fight three Roman legions simultaneously.

The battle over, Crassus faces the thousand survivors seated on the ground as an officer shouts, "I bring a message from your master, Marcus Licinius Crassus, Commander of Italy. By command of his most merciful excellency, your lives are to be spared. Slaves you were, and slaves you remain. But the terrible penalty of crucifixion has been set aside on the single condition that you identify the body or the living person of the slave called Spartacus."

After a long pause, Spartacus stands up to identify himself. Before he can speak, however, Antoninus leaps to his feet and yells, "I am

Spartacus!" Immediately, another man stands and yells, "No, I'm Spartacus!" Then another leaps to his feet and yells, "No, I'm Spartacus!" Within minutes, the whole slave army is on its feet, each man yelling "I'm Spartacus!"

Each man, by standing, was committing himself to death by crucifixion. Yet their loyalty to Spartacus, their leader, was superseded only by their loyalty to the vision of themselves as free men, the vision that Spartacus had inspired. The vision was so compelling that, having once tasted freedom, they willingly chose death over once again submitting to slavery. And they were, to a man, crucified along the road to Rome. But by withholding their obedience from Crassus, they remained free because slavery requires that the oppressed submit their obedience to the oppressor.

In more recent times, a vision of freedom and equality inspired thirteen colonies to formally declare their independence from England on July 4, 1776. The vision of human freedom and equality was so strong that a whole nation, the United States of America, was founded on it. In 1836, the fall of the Alamo, the Franciscan mission in San Antonio, Texas, and the slaughter of the men defending it inspired Texans in their vision of freedom from Mexican rule. In both cases, the strength of the vision carried a people to victory against overwhelming odds.

In contrast to the above example, however, is the movie *Braveheart*, a true story about Sir William Wallace (1272?–1305). In 1296, Edward I, king of England, claiming the Scottish throne for himself, drove out the king of Scotland, stationed English soldiers in the country, and stole the "Stone of Scone," also known as the "Stone of Destiny," which was the ancient symbol of Scottish sovereignty.

William Wallace, known for his courage and strength, led bands of Scottish patriots in a bitter war against the invaders. The English raised an army and advanced against Wallace, only to be defeated at the battle of Stirling Bridge. At that point, Edward hurried home from France and led a great army against the Scottish clansmen, whom he defeated at Falkirk in 1298. Wallace escaped, however, and carried on the fight in the mountains. He was captured seven years later (1305) and executed.[49]

The story is also about Robert The Bruce (1274–1329), who became king of Scotland in 1306 and reigned until 1329. He defeated the British at Bannockburn in 1314 and finally won recognition of Scottish independence from England in 1328.

During the life of William Wallace, however, Robert The Bruce had sworn allegiance to Edward I, king of England, although he occasionally changed sides and aided Wallace.[49] In the movie *Braveheart*, not only did Robert The Bruce change sides but also the greedy noblemen of Scotland changed sides, which meant that Wallace was often betrayed by his own countrymen in the struggle for Scottish independence from England.

The betrayals came about because, in order to defeat Wallace, the king of England needed to weaken the resolve of the vision of Scottish freedom from England that Wallace represented. Edward I would therefore bribe Robert The Bruce and the Scottish noblemen with lands and titles in England, which they accepted—proving the maxim, united we stand, *divided we fall.*

It was precisely this divided loyalty that undermined the vision of freedom held by Wallace and the other Scottish patriots. It was this divided loyalty that caused Robert The Bruce to take another 23 years, until 1328, to finally fulfill the vision of Scottish independence. Had Robert The Bruce and the Scottish noblemen united unconditionally with Wallace and the Scottish patriots behind the vision of freedom, might Wallace have lived to see a free and independent Scotland with Robert The Bruce as king seated on the Stone of Scone (a rough-hewn block of grey sandstone weighing 458 pounds) prior to the year 1300?

They did not unite, however, so it was not until November 14, 1996—700 years after its theft, although it was briefly reclaimed by Scottish nationals (1950–51)—that the Stone of Scone was returned from underneath the Coronation Chair at Westminster Abbey in London, England, to Edinburgh Castle, Scotland.[50] Then, finally, was Scotland's vision of freedom complete.

Although a vision may begin as an intellectual idea, at some point it becomes enshrined in one's heart as a palpable force that defies explanation. It then becomes impossible to turn back, to accept that which was before, because to do so would be to die inside. Few, if any, forces in human affairs are as powerful as a shared vision of the heart. Consider Mahatma Gandhi's inspired fight to free India from British rule.

In its simplest, intellectual form, a shared vision asks: What do we want to create? Why do we want to create it? Beyond that, it becomes the focus and energy to bring forth that which is desired, because, as John F. Kennedy said, "Those who anticipate the future are empowered to create it." Alas, few people know what a vision, goal, or objective is; how to create them; how to state them; or how to use them as guidelines for sustainable development.

A statement of *vision* is a general declaration that describes what a particular person, group of people, agency, or nation is striving for. A vision is like a "vanishing point," the spot on the horizon where the straight, flat road on which you are driving disappears from view over a gentle rise in the distance. As long as you keep that vanishing point in focus as the place you want to go, you are free to take a few side trips down other roads and always know where you are in relation to where you want to go, your vision. It is therefore necessary to have at hand a dictionary and a thesaurus when crafting a vision statement, because it must be as precise as possible; through it, you must say what you mean and mean what you say.

Gifford Pinchot, the first chief of the U.S. Forest Service, had a vision of protected forests that would produce commodities for people in perpetuity. In them he saw the "greatest good for the greatest number in the long run." Through his leadership, he inspired this vision as a core value around which everyone in the new agency could, and did, rally for almost a century.

In a more recent example, Chris spoke in 1989 to a Nation of First Canadians who owned a sawmill in central British Columbia. He had been asked to discuss how a coniferous forest functions, both above- and belowground, so that the First Canadians could better understand the notion of productive sustainability, something they were greatly concerned about. After he spoke, a contingent from the British Columbia provincial government told the First Canadians what they could and could not do in the eyes of the government. The government officials were insensitive at best. The First Canadians tried in vain to tell the officials how they felt about their land and how they were personally being treated. Both explanations fell on deaf ears.

After the meeting was over and the government people left, Chris explained to the First Canadians what a vision is, why it is important, and how to create one. In this case, they already knew in their hearts what they wanted; they had a shared vision, but they could not articulate it in a way that the government people, whose dealings with the First Canadians were strictly intellectual, could understand.

With Chris's help, they committed their feelings to paper as a vision statement for their sawmill in relation to the sustainable capacity of their land and their traditional ways. They were thus able to state their vision in a way that the government officials could understand, and it became their central point in future negotiations.

In another instance, also in 1989, Chris helped a president and vice president frame a vision and goals for their new company. Although the president became frustrated during the two-day process, he told Chris a couple of years later that it had been the most important exercise he had ever been through for his company and that he used it constantly as the company grew.

In contrast to a vision, a *goal* is a general statement of intent that remains until it is achieved, the need for it disappears, or the direction changes. Although a goal is a statement of direction, which may be vague and is not necessarily expected to be accomplished, it does serve to further clarify the vision statement. A goal might be stated as "I will see Timbuktu."

An *objective,* on the other hand, is a specific statement of intended accomplishment. It is attainable, has a reference to time, is observable and measurable, and has an associated cost. The following are additional attributes of an objective: (1) it starts with an action verb; (2) it specifies a single outcome or result to be accomplished; (3) it specifies a date by which the accomplishment is to be completed; (4) it is framed in positive terms; (5) it is as specific and quantitative as possible and thus lends itself to evaluation; (6) it specifies only "what," "where," and "when" and avoids mentioning "why" and "how"; and (7) it is product oriented.

Consider the previous goal: "I will see Timbuktu." Let's now make it into an objective: "I will see Timbuktu on my 21st birthday." The stated objective is action oriented: I will see. It has a single outcome: seeing Timbuktu. It specifies a date, the day of my 21st birthday, and is framed in positive terms: I will. It lends itself to evaluation of whether or not the stated intent has been achieved, and it clearly states "what," "where," and "when." Finally, it is product or outcome oriented: to see a specific place.

As one strives to achieve such an objective, one must accept and remember that one's objective is fixed, as though in concrete, but the plan to achieve the objective must remain flexible and changeable. A common human tendency, however, is to change the objective—devalue it—if it cannot be reached in the chosen way or by the chosen time. It is much easier, it seems, to devalue an objective than it is to change an elaborate plan that has shown it will not achieve the objective as originally conceived.

It is important to understand what is meant by a vision, goal, and objective because collectively they tell us where we are going, the value of getting there, and the probability of success. Too often, however, we "sleeve shop." Sleeve shopping is going into a store to buy a jacket and deciding which jacket you like by the price tag on the sleeve.

The alternative to sleeve shopping is to first determine what you want by the perceived value and purpose of the outcome. Second, you must make the commitment to pay the price, whatever it is. Third, you must determine the price of achieving the outcome. Fourth, you must figure out how to fulfill your commitment—how to pay the price—and make a commitment to keep your commitment. Fifth, you must act on it.

Alexander the Great, the ancient Greek conqueror, provides an excellent example of knowing what one wants and how to achieve it. When he and his troops landed by ship on a foreign shore that he wanted to take, they found themselves badly outnumbered. As the story goes, he sent some men to burn the ships and then ordered his troops to watch the ships burn, after which he told them: "Now we win or die!"

"This is all well and good," you might say, "but how do we go about actually figuring out what our vision, goals, and objectives are?"

Crafting a Vision and Goals

Before you begin to craft your vision and goals, you would be wise to pause for a moment and describe to yourself how you feel about your community: What types of images come to mind? Who do you think about in your community and why? What places do you think about (open space, shopping malls, schools)? Do activities present themselves? If so, which ones? In short, characterize your community, and be sure to do so either by recording your questions and answers on tape or by writing them down.

If you find that you are unsure how you feel about something, take the time to consciously observe your community; see how it functions and how you feel about the way it functions. How friendly is it? How safe do you feel living and moving about in it during the day and at night?

If you are still not sure you have covered all of the bases, put yourself in the position of a consultant who has been hired to characterize your community. What questions would you ask the people? Why did you select these particular questions? What are you hoping they will tell you? Why do you think these particular questions are important? Now continue your observations and answer the questions for yourself.

Based on what you see and feel, what values do you hold that are met in your community and why? Which values are not met and why?

By asking these questions of oneself, it becomes clear that framing good questions is the key to crafting a good vision statement and goals. Use this technique to characterize and design the community of your dreams. What would it be like? Describe in writing its primary elements, remembering that the most important part of community, by the very nature of the concept, revolves around the quality of human relationships and the reciprocal partnership between the community and its landscape.

If even a small group of community members is willing to participate in such a personalized exercise, it would quickly become apparent that the makings of a sound vision and goals are contained in the collective of personal observations, feelings, and values. The group could then craft a "straw" document containing a statement of vision and goals to which the community at large could respond.

Alternatively, a consultant could be hired to design the questions and derive the answers by visiting personally and informally with community members in both their places of business and their homes. Here the watchword is *trust.* The people *must* trust the consultant, because people do not care how much a person knows until they know how much that person genuinely cares about them.

It is, after all, the quality and sustainability of one's own community that are being mapped into the future, and that is no small matter. It is thus important to understand that trust is heightened and the community's purpose is served to the extent that members of the community become actively engaged in the process.

Whichever method is used to gather the information, the people must ultimately craft the vision and goals themselves, with the help of a neutral third-party facilitator. Unless the whole community partakes of the process, a straw document must be produced for the community to comment on.

This straw document is *not* a "buy-in" vote, however, which is no more than a wolf in sheep's clothing. The purpose of a buy-in, which is often used by self-centered governments, agencies, and special interest groups, is to win agreement with a self-serving point of view by convincing members of the community that they cannot reason for themselves and should not try. Instead, they should unquestioningly "trust" those of superior knowledge who *know* what is good for the community as a whole. This approach will not work!

The comment period for the straw document must be long enough to give people the time they require to really consider so important a document. And then, the people must really be listened to, and their comments must be collected, collated, and incorporated into the vision. A vision, to be effective, must be finalized by consensus. And finally, a vision and goals must periodically be reenvisioned to keep them dynamic and relevant in the present to the present and the future, something that is seldom done.

Although it is we who define our vision, goals, and the objectives of achieving them, it is the land that limits our options, and we must keep these limitations firmly in mind. At the same time, we must recognize that they can be viewed either as obstacles in our preferred path or as solid ground on which to build new paths.

Remember, Nature deals in trends over various scales of time. Habitat (food, cover, space, and water) is a common denominator among species; we can use this knowledge to our benefit. Long-term social/environmental sustainability requires that short-term economic goals and objectives be considered within the primacy of environmental postulates and sound long-term ecological goals and objectives.

Each generation must be the conscious keeper of the generation to come—not its judge. It is therefore incumbent upon us, the adults, to prepare the way for the children who must follow. This will entail, among other things, wise and prudent planning, beginning with carefully, purposefully interweaving the very best human values into interpersonal relationships, which are the social glue that holds a community together as it struggles with the notion of sustainable development.

Ask the Children

As we build our shared visions of a sustainable future in which each person's core values and expertise are acknowledged, we must exercise the good sense and humility to ask our children, beginning at least with second and third graders, what they think and how they feel about their future. Consider for a moment that the children must inherit the world and its environment as we adults leave it for them. Our choices, our generosity or greed, our morality or licentiousness will determine the circumstances that must become their reality.

Why, then, do we adults assume that we know what is best for our children, their children, and their children's children when adults as a whole are destroying their world through greed and competitiveness? Why are children never asked what they expect of us as the caretakers and trustees of the world they must inherit? Why are they never asked what they want us to leave them in terms of environmental quality? Why are they never asked what kinds of choices they would like to be able to make when they grow up? (For that matter, why do we not ask our elders where we have come from, how we are repeating history's mistakes, and what we have lost, such as fundamental human values, along the way?)

Where do we, the adults of the world, get the audacity to assume that we know what is good for our children when all over the world they are being abused at home by parents who are not in control of themselves, are being slaughtered in the streets in the egotistical squabbles of adults over everything imaginable, and are being starved to death by adults who use the allocation of food for political gain? We adults do not even know what is good for us. How can we possibly speak for our children?

This lack of responsible care was keenly felt at the June 1992 worldwide Conference on the Environment (Earth Summit), held in Rio de Janeiro. A twelve-year-old girl delivered to the entire delegation a most poignant speech about a child's perspective of the adult's environmental trusteeship. Chris saw a video of the speech in which a child was pleading for a more gentle hand on the environment so that there would be some things of value left for the children of the future. The adult audience was moved to tears—*but not to action*!

Five years after the Earth Summit, with all its promises of attacking global problems, forests are still disappearing, the air is murkier than ever, and the population has increased by almost half a billion people. Thus, the Worldwatch Institute paints another bleak global landscape in its annual *State of the World* report released in January 1997.[51]

Governments are lagging badly in meeting the goals set at the Rio de Janeiro summit, says Worldwatch Institute in its global review, which is distributed in 30 languages. "Unfortunately," says the report, "few governments have even begun the policy changes...needed to put the world on an environmentally sustainable path." Among Worldwatch's gloomiest conclusions are that millions of acres of tropical and temperate forests are still disappearing each year, that emissions of carbon dioxide are at record highs, and that continued growth of the human population is outpacing the production of food.

Worldwatch is toughest on the United States and the World Bank. It says that American leadership has faded since the summit, in contrast to the strides made by Europe in fighting pollution and by Japan in maintaining foreign aid. Eileen B. Claussen, assistant secretary of state overseeing environmental affairs, said Worldwatch's assessment of progress in the United States is "generally correct," noting that Congress not only failed to ratify a biodiversity treaty but also slashed funding for the summit's major initiatives.

With respect to the World Bank, which lends $20 billion a year to poor countries, the report says that while the bank "touts environmental lending," it pours funds into development schemes that both add to carbon emissions and destroy ecosystems.

Not all is doom and gloom, however. The report found hope in growing numbers of grass-roots movements, particularly in Bangladesh and India. In addition, more than 1,500 cities in 51 countries have adopted plans and rules for dealing with pollution that are often more stringent than their national governments proposed at the Rio Earth Summit.

Presaging Worldwatch's assessment of progress (or the lack thereof), Maurice Strong, Earth Summit secretary general, also issued a report in January 1997, which cites pockets of progress but concludes that "far too few countries, companies, institutions, communities and citizens have made the choices and changes needed to advance the goals of sustainable development."[51]

If this is the adults' response to the children's plea at the Rio Earth Summit for a more gentle hand on the environment so there would be some things of value left for the children of the future, one might well ask what the children want. How reasonable is their request?

More than 93,000 schoolchildren wrote letters to President Clinton telling him what they want him to do during his second term in office to improve the world they will inherit. The students participated in a November 1996 survey, "Goals for the President," conducted by *Weekly Reader,* the nation's oldest and most widely circulated newspaper for schoolchildren.[52]

The survey showed that the children's primary goal was stopping violence because the youngsters believe it affects them at home, on the streets, and at school. The other four top priorities were the environment, substance abuse, homelessness and poverty, and education.

"The children are mainly concerned with issues that affect them now," said Sandra R. Maccarone, editor-in-chief of the *Weekly Reader.* "So it's a sad commentary on our society when schoolchildren are concerned about violence and substance abuse—concerns that were not thought about by children, let alone voiced, when many of us were growing up."

With respect to the environment, the children wrote comments like "We want to stop pollution. It makes people sick."

"Please help protect the animals of our world. Don't let them become history," wrote the fourth-grade students of Canyon Creek School in Billings, Montana.

Caitlin Cavanagh, a third-grade student at Seneca Street School in Oneida, New York, wrote: "This is the only planet we can live in, and other things won't matter if we don't have our environment."

Are their requests unreasonable? Or are these simply pleas for social/environmental sustainability? Could the children help us adults to focus clearly on the issue of social/environmental sustainability if we let them?

In the society of the future, it is going to be increasingly important to listen to what the children say because they represent that which is to come. Children have beginners' minds. To them, all things are possible *until* adults with narrow minds, who have forgotten how to dream, who live in fear of one another, put fences around their imaginations.

We adults, on the other hand, too often think we know what the answers should be and can no longer see what they might be. To us, whose imaginations were stifled by parents and schools, things have rigid limits of impossibility. We would do well, therefore, to consider carefully not only what the children say is possible but also what they want. The future, after all, is theirs, which brings us to some of the questions we need to ask as we prepare to engage in sustainable community development.

11

PREPARING TO IMPLEMENT SUSTAINABLE COMMUNITY

To be sustainable, a community must think and act with intelligence, morally sound ethics, and common human decency, which will allow it to work toward the ecological sustainability of its own landscape first and then that of the bioregion. Remember, as you work to implement sustainable development in your own community, that social sustainability may last for a while, even a decade or more, but will in the end collapse if ecological sustainability is not first accounted for, which brings us to the questions that need to be asked.

Questions are a community's most potent futuristic tool when it comes to implementing a vision of a sustainable community within a sustainable landscape. This does not mean, however, that we must know all the answers, which is, of course, impossible. But it does mean that we need to know how to ask a good question.

Because we cannot know all things, we must consciously choose our areas of ignorance and trust that other people are not competing with us for the same areas. If we know well our areas of ignorance, we can use them to help us see and frame questions relevant to the sustainable development of our respective communities.

When recognized and accepted, our personal ignorance and collective ignorance become the precise areas in which we must search for the questions that need to be asked. Viewed in this way, the collective ignorance of a community can be a powerful tool with which to consciously seek answers concerning the sustainability of one direction versus another, of one decision versus another, and for general guidance to an other-centered sustainable future.

WHAT ARE SOME OF THE QUESTIONS THAT NEED TO BE ASKED AS WE PREPARE TO EMBARK ON OUR JOURNEY TOWARD SUSTAINABLE COMMUNITY DEVELOPMENT?

Because old self-centered questions and old self-centered answers have led us into today's disintegration of social/environmental integrity and are leading us toward an even greater loss of integrity tomorrow, it is important to understand that the answer to a problem is only as good as the question and the means used to derive the answer. Before we can arrive at fundamentally new answers, we must be willing to risk asking fundamentally new questions. We must therefore pay particular attention to the questions we ask because the answers we accept will become the consequences of the future. This means that we must look long and hard at where we are headed with respect to the quality of our community, our environment, and the legacy we are leaving the children.

Heretofore our society has been more concerned with *getting* answers to its self-centered questions than it has been with *asking* other-centered (morally right) questions. Politically correct answers validate preconceived economic/political desires. Other-centered questions lead us toward a future in which social/environmental options are left open so that generations to come may define their own ideas of a "quality community" and a "quality environment" from an array of possibilities.

To this end, we must pay vastly more attention to the questions we ask. A good question, one that may be valid for a century or more, is a bridge of continuity among generations. Examples of good questions might include: Are the consequences of this decision reversible, and if so, to what extent? To the degree that the consequences are not reversible, can the apparent short-term social benefits be justified in an other-centered, future-oriented ecological way?

It is, after all, the questions we ask that guide the sustainability of our community/environmental development, and it is the questions we ask that determine the options we bequeath to the future. Asking the right questions, therefore, can create a holistic web of multilevel thinking, which can act as a catalyst for conscious responses to the complexity of human interactions within the organic system of nature, including the community itself.

While it is not within the scope of this book to offer an exhaustive list of questions that need to be asked, we hope it will be helpful to consider the following ones, because the answers we accept will determine whether or not a vision for the future is sustainable: (1) What sources of energy are available to our community? (2) What social capital is available within our community? (3) How can specialities be sustainably fitted into a general community? (4) What is necessary to build a community in an intelligent, moral way? (5) When is enough, enough? (6) Are the consequences of our decisions reversible, and if so, to what extent? (7) How will the things we want to introduce into our community's environment affect its future? (8) How much waste can we convert into food for microorganisms? (9) Will planning benefit us as a community? (10) Why monitor for sustainability?

WHAT SOURCES OF ENERGY ARE AVAILABLE TO OUR COMMUNITY?

Before answering the above question, one must understand a little about the nature of energy. Of the physical principles governing Nature's dynamic balance, three, as far as we know, are inviolate.[53] The first principle, *the law of conservation of mass,* simply states that mass can neither be created nor destroyed; therefore, materials cannot really be "produced" or "consumed." The mass of a material remains the same while its form is altered from a raw material to finished products, wastes, and residuals without a change in quantity. Thus, over time, the amount of matter moving through a stable system must equal the amount stored in it plus the amount moving out of it.

Einstein's "Special Theory of Relativity" states that the energy produced (E) is equal to the amount of equivalent mass converted in a transitory state as the mass of radiant (or other forms of) energy (m) times the square of the speed of light (c, or 3×10^8 mass/speed of light) or $E = mc^2$. This conversion of mass energy to other forms of energy occurs in many processes; for example, it occurs in everything from the energy radiated by an ordinary light bulb, which is too small to measure in most cases, to the measurable mass energy radiated by our sun and other stars.

The second principle, *the law of conservation of energy,* states that energy can neither be created nor destroyed. Thus, while energy can be changed in form and distribution, the quantity remains the same. The notion of either "energy production" or "energy consumption" is therefore a non sequitur.

Energy is neither produced nor consumed; it is only converted from one form to another. For example, when the energy stored in fossil fuels is released ("consumed"), creating thermal, mechanical, or electrical energy, only the form of the energy has changed.

The third principle, *the second law of thermodynamics,* states that the amount of energy in forms available to do useful work can only diminish over time. The loss of available energy thus represents a diminishing capacity to maintain "order," which increases disorder or entropy.

When considering the notion of social/environmental sustainability, it is vital that a community understand and accept (first and foremost) that its energy comes initially from the natural resources of its environment, be it a sustainable supply of timber, fish, or water. Second, a community must understand and accept that to maintain a steady flow of this vital energy, the community must have a healthy environment: healthy soil, clean air, and clean and abundant water for the timber; healthy water catchments, clean air, and clean and abundant water for the fish; healthy water catchments, healthy soil, and clean air for the water. Third, a community must understand and accept that an "expenditure" of energy means the conversion of a useful or available form of energy (that with which work can be done) to a less useful or less available form and must therefore be used wisely. For purposes of sustainable community development, we consider energy in two ways: natural and cultural.

Natural Energy

Current scientific thought says there arose a great cataclysm, the "big bang," which created a supremely harmonious and logical process as a foundation of the evolution of matter, and the Universe was born. So began the process of evolution, which proceeds from the simple to the complex, from the general to the specific, and from the strongly bound to the more weakly bound.

To understand that evolution, consider an extended family. The strongest bond is between a husband and wife, then between the parents and their children. But as the family grows, the bonds between the children and the various aunts and uncles and their first, second, and third cousins become progressively weaker as relationships become more distant with the increasing size of the family.

The big bang created particles of an extremely high state of concentration bound together by almost unimaginably strong forces. From these original micro-units, quarks and electrons were formed. (Scientists propose quarks as the fundamental units of matter.) Quarks combined to form protons and neutrons; protons and neutrons formed atomic nuclei, which were complemented by shells of electrons. Atoms of various weights and complexities could, in some parts of the Universe, combine into chains of molecules and, on suitable planetary surfaces, give birth to life. On Earth, for example, living organisms became ecological systems and human societies with the remarkable features of language, consciousness, free choice, and culture.

In this giant process of evolution, relationships among things are changing continually as complex systems rise from subatomic and atomic particles. In each higher level of complexity and organization we find an increase in the size of the system and a corresponding decrease in the energies holding it together. So as evolution proceeds, the forces that hold together the evolving systems, from a molecule to a human society, weaken as the size of the system increases.

Earth has been exposed for billions of years to a constant flow of energy streaming from the sun and radiating back into space. On Earth, the flow of energy produces a vast variety of living systems, from the simple, such as an individual cell, to the complex, such as a human society. Each system uses the sun's energy to fuel its own internal processes, and each in turn provides fuel to others.

During its evolution, every system must develop the ability to constantly balance the energy it uses to function with the energies available in its environment. Ecosystems and social systems, like organisms, constantly bring in, break down, and use energy not only for repair but also for regeneration and to adapt to changing environmental conditions.

With the above in mind, the necessary questions become: What raw materials, such as water, wood, coal, oil, grazing land, and arable soil, do we have within the vicinity of our community that can act as immediate and/or future sources of energy? How can we sustainably use this energy in the present for the present and the future?

This latter question is critical because some forms of energy, such as nonrenewable fossil fuels, have problems of pollution associated with them when they are converted into other forms of energy, such as electricity. In the process of converting coal to electricity, pollutants, such as sulfur dioxide, are spewed into the air and carried hundreds or thousands of miles from the coal-fired power plants that produced them. This phenomenon of aerial transportation is known as acid deposition, commonly called acid rain, which ultimately affects ecosystems in ways that are deleterious to human survival.

Thus, while coal-fired power plants may offer short-term economic benefits to a segment of society in the present, they are ecologically devastating to the global environment over the longer term and thus are counterproductive to social/environmental sustainability for all generations. There are, however, alternative forms of energy, such as wind power and solar power, that are both clean and sustainable. Although they can be developed, they are currently taking a back seat to the monied interests that are invested in and sell dirty, nonsustainable energy, such as nuclear, oil, and coal. And then there is cultural energy.

Cultural Energy

By cultural energy we mean those forms of energy with which people do work within the community. Cultural energy is "people power," and every community has it.

People's sense of empowerment and belief in their potential to resolve problems are crucial to sustainable community development. People are a powerful catalyst for change both as activists within their local communities and as examples by changing their behaviors to promote more sustainable lifestyles. Such change relies on creativity, which is fragile and easily stifled. Each person's creativity must therefore be encouraged, developed, and protected if sustainability is to be a viable part of community development.

When people are inspired by their own interests and enjoyment, there is a better chance they will explore unlikely paths, take risks, find that employment and income can exist within sustainable practices, and produce something that in the end is useful. Although motivation is internal to people themselves, not everyone is strong enough emotionally to withstand the negative pressures of society to fit into a perceived norm and thereby minimize creating the uncomfortable risk of change. Thus, while people must motivate themselves, they can be helped to analyze, understand, and use their own experiences to new and greater ends. This brings up a Chinese proverb: "I hear and I forget; I see and I remember; I do and I understand." In this sense, people are not trained, although training in formal skills may be valuable; they are liberated and train themselves, using other people as examples and in turn providing examples to still other people.

When goals are imposed on them, however, or when they are goaded by fear of being fired, creativity withers. Sustainable community development therefore depends on intrinsic motivation, which is conducive to creativity.

Although our larger social system is designed to insist on conformity, to go along with mass thinking, local sustainable community development by its very nature is designed as an advocate and protector of the freedom and space necessary for creativity to flourish. The single most important component of creativity is freedom—the power to decide what to do, how to do it, and when to do it, a sense of control over one's own ideas and work.

Because many local issues can and must be addressed simultaneously through the process of sustainable community development, it is a potentially powerful strategy for change. By addressing the needs and concerns of both individuals and groups, the solidarity and adaptability of a local community are increased, which means that the issues on which a community focuses become increasingly centered within the context of long-term sustainability.

For such a strategy to work, however, the following questions must be addressed: In what form (or forms) is a community's people power available? Do we have the kind of people power we need to accomplish our vision? If so, how do we capitalize on it for the good of the community, present and future? If not, how do we increase the capacity of our community's people power—our social capital?

WHAT SOCIAL CAPITAL IS AVAILABLE WITHIN OUR COMMUNITY?

Many a community in trouble makes the mistake of calling in outside "experts" for advice on what it "should" or "should not" do before it has even considered the social capital (collective talents, skills, experiences, and willingness to work = energy) of the people who are the community. Although President Theodore Roosevelt was probably speaking about an individual when he said, "Do what you can with what you have, where you are," the same admonishment applies to a community. And so does Ralph Waldo Emerson's: "This time, like all times, is a very good one, if we but know what to do with it." So, before a community calls for an outside expert, it would be wise to see what it can do for itself with its own resident experts whose talents, skills, experiences, and energy are directly relevant to the needs of the community.

Existing Human Talents, Skills, and Experience

Every community has within it many people with known talents, skills, and experiences, but each community also has among its members many hidden, even latent, talents and many unknown skills and experiences. Community leaders should not, therefore, draw conclusions about the potentialities of the community's membership based solely on past experience. That would be like saying that a dog cannot learn to roll over simply because it has not done so.

While it is of value to recognize what talents, skills, and experiences are already available within a community, it is equally important to realize that people are continually growing and hence so is the pool of talents, skills, and experiences from which a community can draw in the future. Failure to realize people's potential for continual growth calls to mind a comment by poet William Blake: "The hours of folly are measured by the clock; but of wisdom, no clock can measure."

Talent

"Every man," thought Ralph Waldo Emerson "has his own vocation, talent is the call." Talents and skills differ in that one's talent is innate to one's being whereas one's skills are acquired through study, imitation, training, and practice.[54] It is one's talents, not one's skills, that open the door to discovery and innovation, which is not to say that skills are unimportant.

Talents are meant to be shared while abnegating ownership, a lesson Michelangelo learned—and we must too. On the one hand, Michelangelo honored his skill by saying: "If people knew how hard I worked to get my mastery, it wouldn't seem so wonderful after all." On the other, he learned that his talent was a sacred gift to be held in trust.

After he completed his first *pietà,* he heard rumors that another artist was being credited with his work. He thus stole into the cathedral one night and carved his name on the back of his *pietà.* This act bothered him so much that he later called it an abomination and a desecration. This *pietà* was the only piece of work he ever signed because he recognized that his talent was a sacred gift and could not be possessed.[55]

Although vision and its purpose marshal talent, more talent is squandered (both personally and as a community) due to a lack of vision and purpose (the kind Michelangelo had) than from any other cause. Recognizing and developing talent is the road to creativity in both the individual and the community, for a community's creative consciousness is raised each time one of its members experiences inner growth.

If a community is dedicated to a vision and its purpose, members of the community will find a way to express their talents through it. In fact, they might even discover new talents. But remember, each person's talent is equally important to the whole. No one's talent is greater or lesser than anyone else's; they are only different and therefore complementary.

Raw talent is that which is undeveloped. Talent is incomplete, however well developed, when it is not aligned with a vision, purpose, and values. But when talent is integrated with a vision, purpose, and supporting values, there is no telling how far an individual or a community can go, provided the necessary skills are also available.

Skill

While talent is unique to each individual, skill seldom is. Although talent is innate to a person's inner world, skill must be earned in the outer world. More often than not, skill is at least a partial gift from some other person or other people. Nevertheless, we often spend so much time, effort, and money developing our skills that we are loath to admit we may lack talent in the area of endeavor. This is not to say that our skills are wasted, but only that they may be out of harmony with our talents. Nevertheless, each person's skill, whatever it is, is important to sustainable community development.

While some people take their skills for granted or see little value in them outside the immediate necessities of earning a living, others have such low self-esteem that they imagine their skills to be either poorly developed compared to someone else's or of small value to anyone. Yet each person in a community (and her or his respective skill) is a part of the whole and therefore of value to the whole.

Sustainable community development requires the greatest possible participation of community members, and this means all of their assorted skills at some time or another, be they those of a cook, carpenter, electrician, community organizer, facilitator, or custodian. For one's skill to be used to the best advantage, however, one must draw on one's personal experiences.

Experience

Experience is the active participation in events that lead to an accumulation of knowledge or skill, the totality of which represents the past of an individual or group of people, such as a community, archived in language. Every human being holds within the language his or her experiences in and with life, be it a mode of expression carried forward from one's family of origin or the jargon of one's craft or trade. Every human language—the master tool representing the experiences of its own culture—has its unique construct, which determines both its limitations in and its possibilities for expressing myth, emotion, and logic.

As long as we have the maximum diversity of languages and the experiences they represent, we can see ourselves—the collective human creature, the social animal—most clearly and from many points of view in a multitude of social mirrors. And who knows when an idiom of an obscure language or a "primitive" culture, or the serendipitous flash of recognition spurred by the collective experiences of people secreted in some ancient myth or modern metaphor, may be the precise view necessary to resolve some crisis in the implementation of sustainable community development. Here you might ask how leadership fits into the scenario.

Leadership

A true leader is concerned primarily with facilitating someone else's ability to reach his or her potential as a human being by helping that person develop his or her talents and skills and value his or her own experiences. Leadership thus comes from the heart and deals intimately with human values and human dignity, which are the essence of good citizenship. One must lead by example, as Francis Bacon noted when he said, "He that gives good advice, builds with one hand; he that gives good counsel and example, builds with both."

To be a good leader, one must first learn to be a good follower. But too often people in positions of leadership (who lack within themselves the inner authority of true leaders) see themselves in positions of power, which they confuse with leadership and thus with citizenship.

Such people are smitten with power and are loath to relinquish it, even momentarily, for fear of losing it altogether. These people cannot lead because leadership requires a great deal of trust, a clear sense of interdependence, and a clear sense of high principle and the courage to stick with it despite any and all personal costs. Leadership must never be handled carelessly or selfishly.

Servant Leadership

A leader knows and does what is right from moral conviction, usually expressed as enthusiasm, which causes people to want to follow with action. Essentially, a leader is one who values people and helps them transcend their fears so they might be able to act in a manner they were incapable of on their own.

Leadership has to do with authority, which is control, or the right or power to command, enforce laws, exact obedience, determine, or judge, albeit one may lead unobtrusively in the background. Two kinds of authority are embodied in this definition: that of a person and that of a position.

The authority of a person begins as an inner phenomenon. It comes from one's belief in one's higher consciousness, which acts as a guide in life when one listens to it: "As a man thinketh in his heart, so is he."[55] In contrast, a person who has only the authority of position may have a socially accepted seat of power over other people, but power can exist only if people agree to submit their obedience to the authority. A person who holds a position of authority yet does not live from the authority within can only manage or rule as a dictator—through coercion and fear—but cannot lead.

A leader's power to inspire followership comes from a sense of authenticity, because he or she has a vision that is other-centered rather than self-centered. Such a vision springs from strength, those Universal Principles governing all life with justice and equity, as opposed to the relatively weak foundation of selfish desire. It is the authenticity that people respond to, and in responding, they validate their leader's authority.

Managerialship, on the other hand, is generally of the intellect and pays minute attention to detail, to the letter of the law, and to doing the thing "right" even if it is not the "right" thing to do. A manager tends to rely on the external, intellectual promise of new techniques to solve problems and is concerned that all the procedural pieces are properly accounted for; hence the epithet "bean counter."

Good managers who have little or no leadership ability are thus placed at a disadvantage when put in positions of leadership, because all such people can do is rise to their level of incompetence and remain there, in which case an ounce of image is worth a pound of performance. Similarly, a leader with little or no managerial ability, when placed in the position of management, is equally inept because the two positions require vastly different but complementary skills.

An effective leader is the one who guides the process of sustainable community development, and an effective manager is the one who keeps it running smoothly. By way of example, think of driving a herd composed of a hundred head of cattle.

There are three basic positions in driving cattle: point, flank, and drag. The person riding point is the leader, the one out front guiding the herd. The flankers, or people riding along the sides of the herd, manage the herd by keeping it moving in the desired direction while preventing individuals from leaving the herd. Riding drag means bringing up the rear or keeping the cattle moving at a given speed while preventing individuals from dropping out of the herd, which is part of good managerialship. Together, leader and manager are responsible for moving the whole herd safely from one place to another.

A leader must be the servant of the parties involved. Servant leadership offers a unique mix of idealism and pragmatism.

The idealism comes from having chosen to serve one another and some Higher purpose, appealing to a deeply held belief in the dignity of all people and the democratic principle that a leader's power flows from commitment to the well-being of the people. Leaders do not inflict pain, although they often must help their followers to bear it in uncomfortable circumstances, such as compromise. Such leadership is also practical, however, because it has been proven over and over that the only leader whom soldiers will reliably follow when risking their lives in battle is the one whom they feel is both competent and committed to their safety.

A leader's first responsibility, therefore, is to help the participants examine their senses of reality and his or her last responsibility is to say "thank you." In between, one not only must provide and maintain momentum but also must be effective. But beware! Most people confuse effectiveness with efficiency. Effectiveness is doing the right thing, whereas efficiency is doing the thing right, although at times it may not be the right thing to do.

When the difference between effectiveness and efficiency is understood, it is clear that efficiency can be delegated but effectiveness cannot. In terms of leadership, effectiveness is enabling others to reach toward their personal potential through participation in the process of sustainable community development. In so doing, a leader leaves behind a legacy of assets invested in other people.

A leader is also responsible for developing, expressing, and defending the follower's civility and values. Paramount in the process of sustainable community development are good manners, respect for one another, and an appreciation of the way in which we serve one another. In this sense, civility has to do with identifying values as opposed to following some predisposed process formula.

For a participant to lose sight of hope, opportunity, the right to feel needed, and the beauty and novelty of ideas is to die a little each day. For a leader to ignore the dignity of interpersonal relationships, the elegance of simplicity and truth, and the essential responsibility of serving one another is also to die a little each day. In a day when so much energy seems to be spent on mindless conflict, to be a leader is to enjoy the special privileges of complexity, ambiguity, diversity, and challenge.

As auto manufacturer Henry Ford once said: "Coming together is a beginning; keeping together is progress; working together is success." In the end, it is the collective heart of the people that counts; without people, there is no need for leaders. Lao-tzu, the Chinese philosopher, thought a good leader was one who "when his work is done, his aim fulfilled, they will all say, 'We did this ourselves.'" Such is servant leadership.

Revolving or Shared Leadership

Revolving or shared leadership comes about in two ways: first, when "subordinates" break custom and become leaders, and second, when someone's particular expertise is needed and he or she takes over the leadership during that time. Revolving leaders are indispensable in our lives because they take charge in varying degrees as they are needed.

Revolving leadership is the basis of day-to-day expression in the participative democratic process required in sustainable community development. Such participation is both one's opportunity and responsibility to have a say in the future of one's community through the example and accountability of one's personal behavior, by influencing the community's government through participation in the democratic process, and by extending one's willingness to accept ownership in resolving its problems.

Because no one person can be an expert in everything, the person in the official position of overall leadership must have the good sense and grace to support and follow the lead of a person whose expertise is momentarily in demand. It is difficult for many people to be open enough to recognize what is best for their community and to step aside when necessary in favor of issue-oriented or problem-oriented leadership.

In the last analysis, leadership must be shared (but neither given away nor sold) because a time will arise when we must count on someone else's special competence. If we think about the people with whom we share our community, it becomes apparent that we must be able to count on one another if our community is to meet our needs while protecting our deepest values. By ourselves, we are severely limited, but together we can be something truly awesome.

According to Max DePree, CEO of Herman Miller, Inc., a furniture manufacturing company, "The condition of our hearts, the openness of our attitudes, the quality of our competence, the fidelity of our experience—these give vitality to the work experience and meaning of life." Freely and openly shared revolving leadership is one of the vehicles that we can use to help ourselves and one another reach our potential as both human beings and citizens of our respective communities.

"But," you might say, "I'm only one person. What can I do? My actions account for so very little." Because so many people feel this way, it might be instructive to consider snowflakes.

When snowflakes begin falling, those coming down first land on the warm soil and melt, entering the ground without a trace. One after another, they come into view out of the sky, fall past our faces, and land on the ground, only to disappear as rapidly as they appeared—or so it seems.

But each snowflake does something as it touches the soil. It dissipates its coolness into the soil's heat. As flake after snowflake touches the ground and melts, the collective coolness of their beings creates a cumulative effect by which the soil is eventually cooled enough that falling snowflakes melt progressively slower until some don't melt at all. Then snow gradually begins to accumulate until the land is covered in a blanket of white.

Is one snowflake more important than another? Is the one you see sparkling in the sun more important than the one that melted upon landing? Neither is more or less important than the other. Without those that melted and cooled the soil, the ones that ultimately formed the blanket of white would not have survived to do so. Therefore, just as every snowflake (individually and as part of the collective) is important to the whole of winter, so is each person (individually and as part of the collective) important to the whole of a community.

Just as no two snowflakes are exactly alike, no two people are identical. Thus, each individual has a unique gift to offer, a special talent that in the collective of a community is complementary rather than competitive. Each person's belief, being a little different from all the others', helps a community to see itself when that person's voice is raised in expressing his or her particular point of view. And it is the task of leadership to inspire people.

Inspire Performance

Excellent performance can produce optimum results, which may lead to strong participation. A meeting that is well organized and flows easily provides a meaningful, if not enjoyable, time for all and is impressive to visitors who join in future activities.[56]

Behind all of this, as one soon learns, is the hard work of a few people dedicated to the proposition of sustainable community development. A large part of the success of these endeavors is the insistence among those few that they all share in the work. Clear guidelines of acceptable conduct and firm enforcement of those standards will go a long way to ensure equal and meaningful experiences as people strive toward the creation of a sustainable community within the context of a sustainable landscape.

To foster the kind of participation and responsibility necessary to create a successful sustainable community, it is advisable to take seriously the installation of members into positions of leadership within a committee. Elevate the leaders consciously and openly to these positions of prominence and responsibility, and make sure all duties are clear and concise. Although this is especially important for first-time performers, seasoned leaders need reminders of what is expected of them, because negligence is often a matter of not knowing what to expect.

If a problem develops, work to increase the leader's level of commitment. Praise work currently being accomplished and encourage greater excellence by being sensitive to small contributions so they may grow into bigger ones. For example, a leader might say, "Jim is doing a fine job as committee chair for participation by local government, despite his busy schedule. (Applause.) Once he gets his agenda better established, it should be easier to meet with government officials."

At times, one must inquire subtly about work not being completed, but in a way that brings up the problem without criticizing. Remember: *Praise in public, discuss shortcomings in private.* If necessary, have someone check with the delinquent leader before the meeting to see if the work is done. A sincere offer to help as a pretext for the call will avoid humiliation.

If, however, work is still not being done, a more direct approach may be necessary. Call a meeting of the committee leaders prior to or just after the regular meeting, with the individual present. Discuss some normal business and insert a brief discussion about the person's performance toward the end of the meeting—but neither first nor last on the agenda, lest it come across as the main purpose of the meeting.

Because the goal is to call attention to the problem and promote discussion, briefly cite two or three examples of work not being done and express concern for the project as a whole. Allow the person in question to explain his or her difficulties, and honor the explanation no matter what it is. Keep in mind this is a sensitive situation because it is difficult for people to admit they have not fulfilled their agreed-to duty.

Rather than initially asking the person to step down, offer help from a prearranged assistant, who may also be a replacement if the person resigns unexpectedly. One can also provide for an assistant to take over the duties on a temporary basis until the person can find more time. But be sure to get a firm commitment from the individual concerning his or her performance in the future.

If, after all of this, the work is still not getting done and the person is adamant about retaining his or her position, prepare for a replacement, but this must be the last resort. The person must be approached in private by the committee chair and one other officer. State the problem gently yet firmly and ask that a replacement be allowed to assume the duties, which will give the person an opportunity to voluntarily pass forward his or her duties while still retaining as much control as possible of his or her dignity.

If the person refuses to step down, express your concern for his or her feelings, but also express your concern for the feelings of the other members of the committee. It is now appropriate to again ask for the successor to take over, while encouraging the person to try again when his or her circumstances have improved, which continues to allow the person control of the decision and to retain dignity.

If the person insists on continuing, grant one final opportunity by allowing him or her a specific period within which to fulfill the stated duties of the position. If the position is important enough to the person to want to continue, a final chance may be in order, but make it clear that if conditions do not improve, replacement is inevitable.

At this point, the person is no longer in control. If he or she succeeds, the immediate problem is solved; if not, the replacement takes over and the immediate problem is solved. If a replacement must take over, have the incumbent make a brief checklist to familiarize the new person with the duties and to allow private discussion.

Imagine Yourself as Different People

Being a good leader requires one to be aware of many different things happening simultaneously, such as the various ways different people respond to any given circumstance, especially when fear of loss or change is involved. Although conflict often arises where fear resides, a person who can imagine himself or herself in the shoes of another person can often tap into what the other person may be feeling and why. To do this, one must ask, "How would I feel in that person's situation? Why would I feel that way?"

A few people assume this role naturally and automatically; others can learn to do so consciously. A person with this ability can often defuse conflict, either before it starts or before it gets out of hand and requires special resolution. Such a person can also anticipate where events are going and thus help people to have compassion and understanding for and patience with one another along the way. In addition, a person who can put himself or herself in another's shoes can often help people to put their talents (whether specialized or general) together in a complementary (instead of competitive) way.

HOW CAN SPECIALITIES BE SUSTAINABLY FIT INTO A GENERAL COMMUNITY?

We are today in a world of specialities, where one tends to focus narrowly on the bits and pieces of a system rather than the whole of the system itself. The result of such specialization is a disjointedness in how we view our respective communities and their reciprocity with their landscapes.

This is not to say that specialization is bad in and of itself, because it is not. Specialization allows a much greater understanding of the pieces of a system and thus has great value, but *only* as long as people remember where and how to fit that detailed information into the system as a whole, which requires the knowledge and skills of a generalist who is good at synthesis.

To the extent that individuals become rigid specialists, our respective communities become rigidly specialized. This specialization is increasingly apparent through the unfolding impact of the global division of labor and the polarization of global politics. In contrast, a Stone Age village and its neighbors could care for every basic need of its people. Short of such catastrophic disturbances as floods and volcanic eruptions, the Stone Age folk could cope with the vicissitudes of Nature.

Today, however, few if any communities in the United States and fewer than perhaps a dozen societies in the world can produce enough food to supply the necessities of their own population. With respect to societies, the same can be said of energy, water, wood products, transportation, and communication, not to mention the myriad consumer goods most people seem to regard as their birthright. When societies are economically specialized and interdependent, they are more or less at the mercy of other societies for the items they lack.

The increasing specialization in almost all dimensions of contemporary communities makes them more and more vulnerable to collapse from economically powerful competition and/or sudden changes in the environment itself. Specialists are unstable and subject to extinction (unemployment); their usefulness comes and goes, whereas generalists continually improvise, adapt, and adjust to new ways of thinking and doing things.

Modern communities are lured into specialization by the linear thinking of quick monetary gains and material security based on continual growth in population, which is seen as necessary for continual growth in economics, which is seen as necessary for material security. And today's communities are each strongly interdependent on one another as they live side by side and share the commonality of their landscapes, supplies of water, air, systems of transportation and communication, and so on. When we add our ever-increasing environmental pollution and continually exploding human population to this already precarious communal system, it has no choice but to grow increasingly risky with respect to a system-wide collapse.

It is precisely this increasing risk of a system-wide collapse that makes social/environmental sustainability so critical. It is this same risk that puts communities on notice that a generalist's way of continually improvising, adapting, and adjusting to new ways of thinking and acting is the only path of survival. The questions thus become: How do we adapt our community to a generalist's way of thinking and acting? How do we fit specialization into our general community for the benefit of the whole?

The first question is critical not only because that is how a community survives but also because the resulting vision is the path the community has chosen to follow into the future. The second question is critical because, having crafted a vision, the community must now figure out how to incorporate any and all specialization (present and future) into the systemic whole to prevent social/environmental fragmentation and collapse.

There were, for example, many communities in western Oregon that were historically "timber towns," dependent on ancient forests for both their cultural identity and their economic stability (their supply of raw energy). When the forests were overcut and the supply of logs collapsed, many of the timber towns turned to tourism for their livelihood, not understanding that they were simply trading one specialization for another.

In addition, they failed to grasp that tourism (especially "ecotourism") is dependent on the health and beauty of their immediate landscapes, which they had destroyed through the practices of nonsustainable forestry. Furthermore, tourism is just an alternative supply of raw energy and is just as easily abused and just as tenuous as was their first supply—the ancient forests.

It is thus important to a community's survival and the success of its vision that whatever specialization is to occur within the community (present and future) be *complementary* with the vision, not in competition with it. This means that each and every specialization must somehow *add to*, not subtract from, the diversity of the community in a way that is *both* socially and environmentally sustainable. A community can only achieve such sustainability, however, if it is both intelligent and moral in its thinking and behavior.

WHAT IS NECESSARY TO BUILD A COMMUNITY IN AN INTELLIGENT, MORAL WAY?

In a freely democratic community, the cultural gold is separated from the dross. The gold is a community's potential to behave like an intelligent, moral, innovative, and freely democratic organization while on the road to sustainability and beyond. The gold lies in a community's desire for and commitment to planning for the future in such a way that the overall quality of life will continue to be productive, healthy, rewarding, equitable, and environmentally sound. Such planning must represent a philosophical decree of public and private responsibility in the present for the future.

The first step along the way is for a community to identify itself as a community, or as John Dewey said in 1927: "Unless local communal life can be restored, the public cannot adequately resolve its most urgent problem: to find and identify itself."

Community Identity

Who are we now, today? As we have already said, this is a difficult but necessary question for people to deal with if they want to create a vision for the future. The vision they create will be determined first by how they identify themselves as a culture and second by how they identify themselves as a civic organization, which in turn will allow the people to ascertain how best to govern themselves. The concept of citizen government means that citizens not only must possess the skills and dispositions to act as rulers but also must know when this obligation is required. In other words, they must know when the good of the community—present and future—is at stake.

Therefore, the self-held concept of who a people are culturally and how well their community governance represents them is critical to the sustainability of their future. A major problem facing communities today is that people are no longer thought of as people but rather as a group of "publics," which is an amorphous aggregate of individuals and their preferences. In this sense, "public" means whatever aggregate of individuals is being measured at the moment, such as the public as market player, as skier, logger, cattle rancher, consumer, scientist, and so on. But in none of these perspectives is the public thought of as whole persons whose humanity supersedes whatever else they might be.[57]

Thus, how well a community's core values are encompassed in a vision depends first on how well the people understand themselves as a culture, second on how well that understanding is reflected in their self-governance, and third on how clearly it is committed to paper for all to see. Only after people have dealt with who they are, today, can they collectively determine what legacy they want to leave for their children and create a vision with which to accomplish it, because only then do they know.

Visions will vary greatly, depending on how a community is defined. A rural community, for example, will include its immediate landscape and perhaps even its relationship to neighboring communities and the bioregion. Within the "inner city," however, a community may be one square city block and its relationship to the four neighboring blocks facing it. Regardless of how people define their particular community, their success in self-governance depends on their sense of citizenship, which is currently entangled with the term "public."

Community Citizenship

The many schizophrenic splits in the concept of the term "public" are the result of what is missing, according to Manfred Stanley:

> The broad core of the classical approach to citizenship stresses a shared constitution that embodies not only rules but [also] a founding myth, a sense of collective moral history, a common-law tradition, and some conception of a good way of life. The principle of public integration is "civic friendship," a concept designed to call attention to bounds transcending mere commercial or military utility. In this sense of public, policies are designed to affect the fortunes of the commons in a particular, historically contingent, moment of its moral development. This is to say, public policies are initiated and evaluated in light of a mythic vision of...a good society.[57]

Citizenship is not some abstract quality of an isolated individual action. Rather, citizenship is the cumulative effect of actions in relationship, where the whole meaning arises from what might be called their ritual practice. A ritual refers to conscious events that come in regular sequences and acquire meaning from their relation to other events in the sequence.[58]

Outside of their full sequence, the individual elements become lost, their existence imperceptible and nonsensical, and their meaning cloudy. Consider, for example, that the meaning of each day of the week comes from its sequential relationship to all the other days, but only so long as they are in the correct sequence. Mix them up and they lose their meaning.[58]

Citizenship, which is based on mutual civility, recognizes that it is the quality of human relationships that either allows and fosters the sustainability of a community—or kills it. Vinoba Bhave has some clear thoughts on civility as the basis of citizenship:

> I am moved by love. I do not deal in opinions, but only in thought, in which there can be give and take. Thought is not

> walled in or tied down; it can be shared with people of goodwill; we can take their ideas and offer them ours, and in this way thought grows and spreads. This has always been my experience....It is open to everyone to explain their ideas to me...and they are free to make my ideas their own in the same way.
>
> There is nothing so powerful as love and thought; no institution, no government, no "ism," no scripture, no weapon. Love and thought are the only sources of power.[59]

As important as civility and citizenship are, however, sustainable community development is possible only to the extent that people keep learning. One innovative way of learning in a democratic setting is the study circle.[60] A study circle is a small-group discussion format to seek understanding and common ground when people face difficult issues and hard choices. Study circles reflect a growing conviction that collective wisdom resides in groups, that education and understanding go hand in hand, and that learning can truly be available for all.

The circular shape of the study group is important and has its roots in antiquity. In medieval literature, for example, brave knights came from across the land to be considered for membership at the Round Table. King Arthur designed its circular shape to democratically arrange the knights and give each an equal position. When a knight was granted membership at the Round Table, he was guaranteed equal stature with everyone else at the table and a right to be heard with equal voice.

In study circles, participants learn to listen to one another's ideas as different experiences of reality rather than points of debate. Although they may not agree, they learn to accept that, just like blind people feeling the different parts of an elephant, each person is limited by her or his own perspective, which is derived from her or his own experiences in life.

By managing the process themselves, participants engage in the practice of democracy. In a study circle, there is equality, respect for others, and excitement about exchanging ideas. This environment is ideal for people to practice the most fundamental aspects of democracy by reaching conclusions or making decisions through talking, listening, and understanding—through sharing.

Sharing is the central connection in a study circle. Participants are encouraged to act as whole people in that they are not required to separate feelings, values, and/or intuition from intellectual thoughts concerning any topic. They are allowed to think systemically as opposed to being placed in a straitjacket of intellectual isolation.

This sharing as whole individuals allows each person to assume the role of teacher, student, leader, and follower at different times in the study circle, which is critical to the viability of both the democratic process and sustainable community development. Because no one person possesses, with equal skill, all of the talents necessary for the practice of either democracy or sustainable community development, it is vital that individuals learn to accept and share the many facets of their personalities to the best of their ability.

People seldom partake of study circles just to learn the so-called objective facts; rather, a study circle deals with real daily problems in the lives of the participants and is thus education in and for life. It is imperative that what the participants learn is grounded in their own experiences and in the real problems and issues they face daily.

Study circles bring people together to talk and to listen; to act and feel as if they are part of a community; and to practice equality, acceptance of ideas and diversity of people and points of view, democracy, and the connectedness of sharing. If an increasing number of people became involved in study circles, it might become clear that the apparent apathy that Americans exhibit toward education and participation in politics is really a disguise for a deep hunger to learn within the safety and nurturance of community.

We say this because, as Myles Horton expressed it: "The fact is that people have within themselves the seeds of greatness, if they're developed. It's not a matter of trying to fill up people, but to fulfill people." "This is all good and well," you might say, "but how can we fulfill people?" We think the answer lies in helping communities become intelligent, moral organizations with a vision of social/environmental sustainability to be enjoyed in the present and passed forward as an unconditional gift for the generations of the future.

Community as an Intelligent, Moral Organization

Our American society has a peculiar split personality with respect to the issue of centralized power. We openly criticize the government-controlled economies and news agencies of other nations because we believe that centralized power leads to the suppression of liberty and bureaucratic corruption and waste. But most of our government bureaucracies and private institutions operate under the assumption that internal centralization of power leads to efficiency and that free choice creates and maintains inefficiency.[61]

Efficient is defined in the *American Heritage Dictionary* as "acting or producing effectively with a minimum of waste, expense, or unnecessary effort." Efficiency is defined (in the same dictionary) as "the ratio of the effective or useful output to the total input in any system; especially, the ratio of the energy delivered by a machine to the energy supplied for its operation." Because individual rights and the freedoms they protect invite inefficiency, power is centralized under the guise of efficiency to omit the human dimension whenever possible and, with it, the democratic process.

The question thus becomes: How do we as a society build democracy back into our communities? To examine this question, we will borrow heavily in the following discussion from some articles and a book by a farsighted husband and wife team, Gifford and Elizabeth Pinchot, whom we acknowledge with gratitude.[61,62] While the Pinchots deal with corporations as intelligent organizations, we have adapted their work to local communities as intelligent, moral organizations. We begin by discussing the distribution of power.

Balancing Power

Because absolute power corrupts absolutely and effectively squanders the human spirit and kills the human will, every revolution has the same theme—balancing the power by flattening the pyramid of the controlling hierarchy. To keep our governing bureaucracies flat, especially those that affect communities, and to make the most of the flattened structure, we must have tools that reach far beyond dehumanizing efficiency. For communities struggling toward social/environmental sustainability, these tools include empowering education for all and the rights of free speech, of assembly, of democratic decision making, and of joint and private ownership, but with moral provisions to protect the ecological and personal rights of the generations of the future.

Many revolutions have a common denominator, which is a basic shift of the day-to-day control and feedback systems from the enthroned hierarchy of power brokers to committees of people. The current revolution of quality (as opposed to quantity) is the most successful attack on overweening bureaucracy, and it is has inspired much progress in developing systems of work, including communities, that are both democratic and collaborative.

In a community, it begins with the recognition and acceptance that the quality of intelligence needed to meet the requirements of social/environmental sustainability demands the expertise and creativity of every person. This means teaching members of the community the skills necessary to ask relevant questions, seek answers, make democratic decisions, and then implement them.

As soon as the common citizens of a community empower themselves to make decisions in interdisciplinary committees, the power vested in the controlling hierarchy is challenged and the belief in the efficiency of specialization is undermined because the teamwork of a dedicated committee remains focused on and acts on the issue before it. What makes this work is the ethical underpinnings of the committee's other-centeredness, which directly contradicts the self-centeredness of the controlling hierarchy.

The success of a sustainable community as an intelligent organization depends on its sense of and commitment to an ethical foundation for its day-to-day operations. Ethics is no longer a luxury, because in these times of growing limits of renewable natural resources on a global basis, ethics is the necessary staple in any sustainable human endeavor.

There are no more plentiful, cheap resources to solve our problems, and continual growth as a panacea no longer works. We are now faced with a suite of dilemmas: Do we need more freedom or more control? Do we need more soft humanism and equality or more hard-line discipline and sacrifice? More collective cooperation and coordination or more individualistic risk and initiative?

The bureaucratic solution of more control (more hard-line discipline and sacrifice, more individualistic risk and initiative, which then needs to be controlled by the hierarchy) no longer works. We need more open, self-organizing communities, modeled on Nature, which have the capacity for self-renewal in a way that is more than simple self-preservation.

The inner security encompassed in the capacity for self-renewal is why old-fashioned values are having a renewed following. We are rediscovering (relearning) that we only bring our best selves to the table when each person is valued for what he or she has become as an inner person, for his or her contributive talents, and when there is sufficient freedom and safety to share the personal gift he or she has to offer. This in turn is either nurtured and protected by the ethics of individual conduct in the collective of a local community—or it is destroyed.

Ethics

Freedom, in turn, is real only when it is exercised with a state of mind well beyond the limits of self-centered individualism. A self-organizing human community depends on each person within the community behaving in an ethical manner, even when no one is watching and there are no penalties for misbehavior. For trust to grow and manifest itself as cooperation, coordination, and adaptability, that trust must be based first and foremost on faith and second on an assurance of the goodness of others within the community.

For a community to be sustainable, it must be organized in such a way that freedom and personal initiative can cohabit with cooperation, coordination, compassion, and a highly integrated harmony. For a community to be sustainable, it must be grounded in the ethical basics of freedom and democracy, which include recognizing, understanding, and accepting the value of the following: (1) diversity; (2) decentralized power; (3) shared and revolving leadership; (4) continuous self-testing; (5) local, regional, national, and global ethics; (6) acting for future generations; (7) making sure that jobs and economic opportunities are available for young people; and (8) practicing the Golden Rule.

1. Diversity—For freedom to be real and collective, it must encompass, value, nourish, and protect diversity. This means that the uniqueness of each and every person, expressed collectively as diversity, is inherently of equal value, is an open-ended potential for contribution, and is equally necessary to the sustainability of the whole.

Within each ethnic and cultural group there resides an enormous diversity that individuals bring into a community—diverse experiences and therefore frames of reference, styles of organization, talents, and competencies. Freedom, not only to be diverse but also to be valued for it, is the leavening of dignity and effectiveness within a democracy. Assigning equal value to each of us brings out our uniqueness and drives the success of interdisciplinary committees, which are the foundation stones of sustainable community development.

2. Decentralized power—For a community to be flexible, responsive, and adaptable, intelligence and creativity must be distributed throughout it, which means that each and every person must be encouraged to use his or her mind and ingenuity and must be openly valued for doing so. To be effective, people must interact in such a way that new ideas and information are shared and used continuously and rapidly. It is important to remember that there are only good ideas, some of which will not work, but there are no bad ideas. To label an idea as bad steals dignity from the person offering it and may cause that person and others to withdraw.

To achieve such flexible, responsive, and adaptable intelligence and creativity, the power to decide and act must be distributed throughout the community. The forces for maintaining centralized power in practice are so great that for a community to be effective, it must raise the sharing of power to an ethical principle, backed up with some level of control through recognized authority.

In practice, the most viable communities are promoting the emergence of new, informal, interdisciplinary committees to develop as necessary across all traditional boundaries within and outside of the community in a concerted effort for the community to become sustainable. This activity replaces the simplistic hierarchy with an amorphous and fluctuating complexity of ever-adjusting relationships.

When the members of a community are increasingly self-empowered and self-organized, simple rules and rigid policies are insufficient to guide them. There must be a shared vision of where they want to go and what they want to accomplish embedded in a deep respect for the ethical foundation on which they stand, because the ends, however nobly perceived, never justify unethical means. The authenticity of leadership is providing ethical and effective power to the people.

3. Shared and revolving leadership—Authentic leadership must be as widely shared within a community as are intelligence and creativity. Free people can only cooperate, coordinate, and act with intelligence if they share a vision toward which to build, in addition to which they must have the ability to test and to renew themselves with accurate feedback. People find it easiest to trust one another when they are working actively toward a shared vision based on a common and widespread moral ethic.

Shared leadership integrates the paradoxes inherent in human relationships. New organizational designs must encourage equal measures of freedom, cooperation, and coordination with equal doses of fiscal discipline. Like all paradoxes, the opposites must be unified to be whole and benign. Anyone who ignores one side of a paradox will be burned by it. We must therefore develop in ourselves both the analytical and the intuitive, the shrewd trader and the compassionate compatriot, the entrepreneur and the communitarian, the individualist and the egalitarian, the devil's advocate and the partner. Wholeness, which requires constant testing, is the hub around which sustainability revolves.

4. Continuous self-testing—Communities that are energized with self-organizing committees and projects, that are flexible and responsive, gain needed cooperation and coordination from continuous self-testing against broad ethical principles and a shared vision. To give people freedom and power is one thing; to assure their ability to use it wisely is quite another.

To examine the wisdom with which members of a community, and therefore the community itself, use their freedom and power requires self-testing against the shared vision and the ethics of its foundation. Such testing not only must occur continuously throughout the entire community but also must promote truth and free and unlimited access to information.

5. Ethics on a global scale—Ethics must be like ripples in a pond—ever expanding. While freedom and power extended to people imbued with a deep sense of local community must initially inspire the ethics of community, ethics must grow beyond the parochial definition of what the individual is initially part of. In other words, ethics must ultimately grow beyond a local community if the community is to become a free, empowered, and responsible member of the bioregion, nation, and world.

When a local community extends its ethical sphere beyond itself, it achieves a kind of aliveness that is self-reinforcing and contagious. The community is teaching by example.

6. Acting for future generations—We must understand, accept, and be accountable for the circumstances of the future because they will always be rooted in our decisions and behaviors in the present. The most difficult and important challenge of our times, therefore, is to discover and create ways in which social/environmental sustainability can be achieved. To accomplish this, we must work diligently and honestly to determine, as best we can, both the cultural capacity and carrying capacity of each local community, county, state, bioregion, nation, and ultimately the world.

We can only accomplish this goal by protecting the options for the generations of the future and by being mindful of how we behave today. Put simply, our choices and actions set in motion forces that create the circumstances the children of tomorrow will inherit as the legacy of consequences we have left for them. The question we must ask ourselves is: Would we want to—or could we even—contend viably with the circumstances that we are leaving for the generations of the 21st century, such as increasingly polluted air and water and growing uncertainty and insecurity in the job market?

7. Making sure that jobs and economic opportunities are available for young people—The most difficult and important social/technological challenge of our time is to find ways to bring the whole human population to a quality of living that nurtures human welfare and honors human dignity while simultaneously protecting the sustainability of the environment. This cannot be done with existing technology, much of which not only is environmentally dangerous but also is contrived to eliminate people's jobs.

Those corporations that value and nurture their people, because they see in them the real corporate wealth, will have the competitive edge in the future. The competitive edge in this case is the loyalty and imagination that each person brings to his or her job. The competitive edge includes sound ecological principles applied to industrial goals and practices not only because they will save money in the face of inevitable regulations and taxes to come but also because good public relations is good marketing.

Wise corporate management will therefore ensure that young people have quality jobs and sound economic opportunities if they want to maintain their competitive edge in the future. Downsizing is anti-competitive over time because it squanders a corporation's real wealth—the loyalty, imagination, and the healthy long-term interpersonal relationships of its people, which brings us to the Golden Rule.

8. Practicing the Golden Rule—"Do unto others as you would have others do unto you." This statement not only seems simple enough but also is a cardinal rule of local community life and survival. Any community that is sustainable has achieved its harmony by treating its own members, as well as it neighbors, as equals, with compassion, dignity, and respect.

The Golden Rule is needed internally in the local community to assure peaceful discourse, freedom and empowerment, cooperation and coordination freely given, and a sense of unity through compassion and mutual caring. The Golden Rule is the ethical pivot of life's compass on which the behavioral needle swings in seeking its direction, and the direction chosen determines the lasting value of that which is done.

These days, it seems that what we call the "realities of life" are harsh indeed and getting harsher. This harshness can be softened, however, when we know we can count on others and they on us, which is the underlying value of a local community.

In some deep inner recess, everyone knows that a renewed commitment to personal ethics; a shared vision; and farsighted, other-centered behavior are necessary to revive the real, personal sense of local community. Many of today's problems result from moral lapses in self-discipline and personal accountability for one's own behavior, as well as materialistic, self-centered overindulgence. To prevent these problems from becoming an untenable burden for the generations of the future, we must all have an unshakable desire and commitment to discern that which is truly ethical and apply it both within ourselves in secret and among others in public.

To do this, we must find people who are content to be servant leaders, people who can help their followers instill within themselves the ethical awareness necessary to be other-centered by sharing their creativity, intelligence, competency, and love for the sustainable good of their local community. Such leaders must foster intelligence and honor intuition among members of the community, especially those who are willing to serve as volunteers on local committees.

Intelligence and Intuition

How can the intelligence and creativity of every member in a community be liberated? How can the ideas, inspiration or intuition, and intellectual analysis of members be integrated into both rapid decisions and actions for the sustainable good of the local community?

No centrally conceived design is capable of producing or allowing the freedom necessary to empower individual intelligence and honor individual intuition while simultaneously knitting together free thinkers in coordinated action with a single focus. Organizations that simultaneously foster intelligence and honor intuition must grow out of the convergence of processes within a community. For this to happen, however, seven conditions must be established based on freedom of choice and democratic participation: (1) widespread truth, evaluation, and rights; (2) liberated employees and committees; (3) freedom to be enterprising; (4) justice and equality; (5) processes for self-management; (6) voluntary networking; and (7) limited community government.

1. Widespread truth, evaluation, and rights—People can only make responsible choices if they know what is going on. Whereas self-centered bureaucrats tend to hoard information, thinking it their source of personal power, a community must create a free exchange of information if it is to act with intelligence.

The Central Intelligence Agency (CIA), for example, is the antithesis of an intelligent organization. It is instead a body designed to collect information in secret, analyze it in secret, decide in secret who should know what, and parcel the information out piecemeal in such a way that the agency's secrecy is not only retained as its source of power but also is constantly growing.

If, therefore, a local community is going to act as an intelligent organization, it must make available full financial information and train all community employees, and any member of the community who wishes to know, how to read the financial statements. All activities that in any way affect the common good of the community must be regularly evaluated and those evaluations immediately posted in the local newspaper or aired on the local TV station for all who are interested to read and/or hear. It must be safe to openly discuss problems that arise and any strategic options to resolve them. There must be a constant iteration and reiteration of how each part fits into the whole and how the whole requires and cares for each part. Finally, freedom of speech and freedom of the press are essential if a community is going to act with intelligence, as is the right of inquiry and of learning for the betterment of the whole, present and future.

2. Liberated employees and committees—Behind nearly every recent innovation in sustainable community development is the superior effectiveness of team-oriented employees and team-oriented committees. Team-oriented committees are the basic building blocks of a community as an intelligent organization.

To reap the benefits of autonomous, empowered committees, we will need the following: (1) a committee must have meaningful authority; (2) it must have free choice of its task, partners, members, and connections; (3) each committee must be evaluated and rewarded as a whole; (4) the people must be trained in the processes of self-management and systems thinking; (5) cooperation and coordination must come from within the committees themselves rather than from a supervisory level downward; and (6) the committee's purpose must be worthwhile in and of itself and must be integrated within a larger vision, the achievement of which is of value to the members of the committee as a whole.

3. Freedom to be enterprising—Intelligent communities release the creative energy of individuals and committees both by making it safe to be creative and by preventing monopolies of power from squelching them.

4. Justice and equality—The kind of democracy needed for members to design and redesign their own community must make creativity and intuition both welcome and safe and must encourage their expression through direct participation in a free democratic process. Liberated members of a local community must be trusted to have the sustainable good of the community at heart.

Communities designed to bring out the responsibility, intelligence, and creativity of every member will depend on internal systems for guaranteeing justice and equality. These internal systems must protect the people from imbalances in power and use sound processes of transformative facilitation to resolve conflicts.

Communities as intelligent organizations will grant freedom that is limited by clearly stated internal laws and an effective system of justice, rather than depending on bureaucratic supervision to prevent abuses of power. The results are better control and, within that, more freedom to be creative.

5. Processes for self-management—Before people are likely to become involved in a self-managing committee, they must know that there are processes in place to support them. When these processes are effective, they engage people in collaboratively managing the whole, which means that all voices are heard and respected so that innovation can take place. This results in the implementation of more new ideas and fewer unthinking mistakes.

Intelligent communities involve all employees and members of volunteer committees in creating the larger context for their work. Consent and consensus guide, as much as possible, the design of the policies and institutions necessary to steer the community toward fulfilling its vision. The result is systems, strategies, and policies that respect the needs of people to feel productive and to continually adapt to changing conditions by experimenting with new ideas.

6. Voluntary networking—For a community to be flexible and responsive to changing conditions, intelligence and intuitive insights must flow from every involved member of the community. Every person must interact in such a way that new information is rapidly disseminated and applied. Only voluntary networking can forge the linkages necessary for such massive, flexible, interconnected conduits of information, because they must be created moment by moment through the choices people make in establishing the connections they need to accomplish their respective tasks.

7. Limited community government—For any society (including a local community) to exist, it must have some form of government to ensure personal rights, safety, and other basic requirements of the common good. The central government of an intelligent community is therefore limited in scope and power because its primary role is to create and protect conditions that allow self-empowered members of the community, employees and volunteers alike, to build the systems necessary to guide the community toward its shared vision.

Status

One of the tasks of leadership is to raise people above selfish, parochial concerns by inspiring them to work together by example for the common good of the whole. The hallmark of good leadership is a community so focused on its shared vision that the members have little desire or energy to squabble over status. A leader can help people satisfy their need for recognition in at least four productive ways: (1) make belonging to the group a source of pride, (2) recognize inner growth and self-control, (3) spread honest recognition throughout the group, and (4) move beyond postures of dominance and submission.

1. Make belonging to the group a source of pride—Leaders inspire their followership by focusing on the shared vision with such authentic enthusiasm that people *want* to join in the adventure of achievement. They lead the group in a constant celebration of its being and its moral growth, as well as its material achievements. A leader sets both ethical and material standards of structure and excellence by example, so that every member can see a path to achievement. By helping the group see itself in the light of its own collective uniqueness and skills of both revolving followership and leadership, the issue of relative status within the group is diffused.

2. Recognize inner growth and self-control—By rewarding people for the most difficult thing any of us ever do, which is to grow through self-discipline beyond where we feel comfortable, people are valued for what they are and are becoming in the best sense of themselves. What greater value is there for a person than the inner growth that leads to a greater sense of consciousness and the personal freedom that comes with it? What greater value is there for a community than to have as its member a person of such courage?

3. Spread honest recognition throughout the group—Sensitive leaders want all the people they serve to receive recognition when it is justly earned, for only then does it have real value. Recognition or status is variable within any system, but it can be generally increased by seeing that the potential of people is released in settings where each individual is valued equally as a person first and foremost. Beyond that, individual recognition is forthcoming based on how a person grows and serves the community as a whole.

4. Move beyond postures of dominance and submission—One of the ways we humans read status is by noting dominance and submission in others and feeling it in our own interrelationships, just as other social animals do, such as monkeys and wolves. Centralized governmental power within communities taps into relationships based on dominance and submissiveness as a matter of control, which means a perpetual struggle for dominance within community government, where the "successful" ride others' fear of failure to the top by making those around them frightened and submissive. In turn, others ride behind the truly fearsome by identifying with reflected fear to intimidate those of lower rank.

Such behavior is grievously flawed, however, as a basis of an intelligent community because those on top are isolated from what is going on beneath them. Isolation leads them to believe their own self-serving version of reality, and no one dares contradict them. The resulting forced submission brings out resentment and its accompanying resistance, which destroys creativity, initiative, self-esteem, and sound choice. There are, however, some ways around the either/or relationship of dominance versus submission:

1. To become fully functional adults, we must each find our own place in relationships that are beyond both dominance and submission; for example, the friendship of equals, the working partnership of equals, the temporary relationship of guest and host, none of which is based on dominance and submission.
2. By having the courage to design human systems that lead toward relationships of equals, a community leader can encourage people to give up their submissive behavior and accept the responsibility of self-guided action.
3. Move toward the revolving status of required expertise, which necessitates giving up the status of a "pecking order." Revolving status of expertise means that when a given person's expertise is required, he or she automatically assumes temporary leadership until someone else's expertise is needed, in which case she or he assumes temporary leadership, and so on. In revolving status, the unspoken rule is that I am dominant in my area of expertise based on my greater knowledge and you in yours based on your greater knowledge; I will recognize your dominance when appropriate and you in turn will recognize mine.

   With time and personal growth, it even becomes possible to move beyond the need for the status of expertise as people move naturally into the next step of interdependence, which brings more activity into the mutual space between the territories of expertise. This is the domain of equality and intelligent interaction from which partnerships are formed.

Partnership is a relationship that is beyond the concerns of relative status. In partnership, the issue is the smoothest function of the whole for the common good, where each person cares for the other as an end, not as a means. By faithfully looking out for one another's dignity and interests, the partners build mutual trust that allows both cooperation and coordination, as well as sharing resources.

4. One advantage of a free democracy is that it allows the growth of a self-organizing human system constructed on the freedom of choice, which expresses itself in the voluntary associations of people from all walks of life who share a common vision toward which they are building. A system of many choices works because a decentralized democratic system of government, for all its faults, is more egalitarian than is a centralized hierarchy as a system of government. It provides a firmer foundation for relationships based on choice rather than ones founded on compulsion.
5. Leaders of intelligent communities tend to select symbols that blend in and make them accessible, such as common parking lots and modest dress. The intelligence of a community can be increased by reducing the differences in status and thereby focusing people's efforts on the community's shared vision, rather than on gaining power over others.
6. Create an array of sincere rewards for many winners, which means focusing attention on areas where everyone can win—as opposed to salaries.
7. Give recognition to those who use it wisely, so the status they achieve can be replicated. Pay close attention to which behaviors are being rewarded and see that good, morally honest behaviors win.
8. Treat everyone with dignity and respect. People with a good sense of dignity and self-respect are less influenced by how others view them and thus are less vulnerable to the blandishments of any system that allocates external status. Wise leaders conserve, create, and apportion recognition with the same care they devote to ethical conduct and responsible fiscal management. They build on voluntary agreements among equals within

self-organizing human systems, which brings us to the notion of moving beyond bureaucracy.

Beyond Bureaucracy

When decisions are made from above and passed down, they affect things that people at lower levels in the hierarchy know about, but these people are neither asked for their information nor listened to when they offer it. They therefore stand idly by watching their community do foolish and wasteful things. They become inured to the decision makers' lack of focus in the relatively small daily decisions, which cumulatively affect the larger decisions the decision makers think they see with such clarity.

The way out of this closed box is to install leaders that will replace blaming and quick fixes with a focus on discovering the dynamic relationships within the community and then open the community to changes that permit the widespread use of intelligence, intuition, and creativity. This means that the community must begin to live on a free and open exchange of information, including free speech and free press, and anything that blocks the flow must be viewed with suspicion.

Truth must be a community's primary value, so that no thought will be silenced or punished. Everyone must be free to comment; any manager who attempts to silence unfavorable comments must be removed.

Intelligent communities have many layers of self-organizing human systems, each of which is interrelated to the others. Consider, for example, the human body, in which the intelligence to form a cell is found within each and every cell, not in the brain. And even in the brain, intelligence is an emergent property of the interaction among the cells of the brain, and not the sole possession of any particular cell. Self-organizing communities, like the brain, require too many subtle connections for any one person or committee to design. The most intelligent communities are designed moment to moment by people interacting freely to get done the things they know need doing.

Whenever feasible, an intelligent community opts for the freedom of choice, as opposed to a monopoly of power. Choice may appear to be less efficient than a monopoly in the short term, but over time, freedom of choice leads to self-motivation and continual innovation, which increases effectiveness as well as cost efficiency. It also allows people to make mistakes together, live with the consequences, and clean up the mess without recrimination from above.

The best communities are therefore designed to give nearly everyone control over something so that everyone has sufficient choices about themselves that they take charge of their own direction and effectiveness. An intelligent community is, in effect, a community of free people who feel safe enough in both their community and their own competence to form groups through which they cooperate and coordinate with a minimum of egotistical turf wars.

This is possible because when a given person's expertise is required, he or she automatically assumes temporary leadership until someone else's expertise is needed and she or he assumes temporary leadership, and so on. The paradox is that to get people beyond defending turf, they must be allowed to have turf within acceptable social constraints.

Intelligent communities can only be built with trust and the freedom trust endows, which means behavior must be ethical at all times. To focus on ethical behavior is to focus on a worthwhile purpose. The higher the purpose, the greater the ethical integrity and the greater the community's intelligence. Once a community's ethical integrity and intelligence are raised in proportion to the quality of its purpose, all incongruous systems within the community, meaning those based on lower-level purposes, will be replaced automatically and rapidly, because within the domain of purpose, higher ones tend to drive out lower ones.

Just as domesticated dogs behave like puppies all their lives when compared to wolves, so too have many of us sold our dignity as adults for the steady wage. Therefore, communities that learn and apply the secrets of freedom, voluntary interdependence, and the co-responsibility of community will attain and continually develop intelligence, intuition, and creativity.

The exemplary symbol for the beginning of an intelligent organization is the lever above the head of every assembly-line employee at Toyota. This lever gives the employees the power to stop the assembly line if something is wrong. Inherent in this lever is a great leap of faith, which is the beginning of a revolutionary shift from trusting in the power of bosses to trusting in a well-designed human system with built-in freedom that honors the integrity and intelligence of the ordinary worker.

Freedom and meaningful communication are difficult to find in our current human world. Any community that incorporates more of either will reap accruing dividends and help to reshape communities in the next century. But in addition to the human component, there is also the sense of place inspired by and through the feeling of one's community within the context of its surrounding environment.

Community Within a Sense of Place

In speaking of community as a sense of place, we do not in this case mean the physical location in which a particular community happens to rest, but rather the human/spiritual component of community as expressed most clearly in the monastic concept of *stabilitas* or stability.[63] Simply put, stability in community implies not only abiding in a particular place but also identifying oneself with the community in all its works, its ups and downs, its tensions, joy, and sorrows. Stability in this case means perseverance in and with a community over the long term for its common good.

In a monastic sense, the vow of stability is based on the fact that a monk, under the appearance of greater potential, which we might call a "greener pasture," may follow a path from one community to another and in so doing lose the good already at hand. The purpose of the vow is to make a monk (man or woman) realize that stability in and of itself is an immense good and that in a vast majority of cases constitutes a much greater good than might be gleaned by changing communities.

If a monk will retain stability, he or she will be able to effect the greatest and most important change—that of oneself, the inner transformation into a more conscious human being open to the balance between spirituality and materiality. If we continually seek outer distraction by moving from community to community, we will continually lack the focus to achieve the inner transformation that could be ours here and now in this community.

In monastic life, it is not sufficient to remain in the same place. It is also necessary to be under the direction of a spiritual teacher. In a secular community, this might be translated to mean that it is not sufficient to remain in the same place; it is also necessary to be committed to a high communal purpose. Thus comes stability in commitment.

Community is for the people who make it up. Stability, therefore, is not primarily for the perfecting of the community in the sense of making it more stable and orderly. Rather, stability is to root the people in the search for a common humanity in which such things as love, faith, trust, mercy, compassion, sharing, and justice become the human foundation of democratic governance.

By committing ourselves to become members of a community, we are above all making a promise of stability, in which we take upon ourselves the responsibility of offering our help to other members of our chosen community, both in the material and in the spiritual. The commitment of stability is not only a promise to persevere within the physical community but also, and most importantly, a promise to persevere in our commitment to live as best we can the love, faith, trust, mercy, compassion, sharing, and justice that make our community a place of safety and personal value.

A community may be a safe place, but to hold people within it and to bring people back after they have experienced more of the world, a community must have things to offer that make staying and/or returning a desirable option. In large part, sustainable community development is creating just such value, which means there is a better chance that those who do leave will come back, bringing with them ideas and skills necessary for the betterment of the community as a whole. This is the hero or heroine's journey described by mythologist Joseph Campbell in *The Hero with a Thousand Faces*.[64]

In this journey, a youth travels away from his or her village to explore the world. Having found adulthood, expressed as some sense of self-mastery, the man or woman returns home and teaches the villagers through service what he or she has learned. But there must be a compelling reason for youth to return as adults.

The higher the moral purpose of a community's vision, the greater its commitment to sustainable development, the more of an intelligent organization it becomes, and the more value the community has to offer those young members who will in time become its leaders. Thus, the more personal the value one feels about one's community, the easier it is to identify one's community with a sense of place, which, after all, is the foundation of sustainable community development. And it is this very sense of value that one wants to keep that brings up the question of how much personal and/or community materialism is enough.

WHEN IS ENOUGH, ENOUGH?

When is enough of something enough? This seems like a simple question. When you have eaten your fill, for example, you quit eating because you know you have had enough, for the moment at least. But what about "enoughness" in the material sense, other than being immediately satiated with food?

Although often referred to as a "standard of living," material enoughness really has to do with our sense of survival through competition, which is usually based on a disaster mentality of impending doom. After all, when one's credo is "more is better," then enough never comes.

The questions we are asking now are unwise questions because they are based on greed. To date, for example, we ask: What is the absolute maximum that we can get out of Nature, people, or a business? What is the absolute minimum that we must legally leave behind for whatever reason?

For example, two small airlines in Oregon both had the opportunity to be sustainable in business, but neither knew when it had enough. The first of these airlines had an excellent business. It seemed to be booked to capacity all of the time. Then the airline decided that it needed to grow, and so it added a route between Oregon and Idaho, where it had not flown before. That extra route caused the company to go bankrupt within a year.

The second airline started with one plane and began immediately building a business. But instead of waiting until it had at least a sustainable business with its one plane, it added a second plane and was bankrupt within a year. In both cases, these companies could have been sustainable if they had known how to pace their growth and when to consciously stop growing—when enough was enough.

On the other hand, consider a lumber company in California that owned land with redwood trees on it. The company was in fact sustainable, had been for nearly a century, and planned to be so for centuries to come. It cut and milled its timber at a rate that would allow its planted trees to grow for 300 years before they would be harvested. The company treated it employees as partners and cared for them and their children as though they really mattered, which they did.

Then, because the company had become so valuable financially, due in large measure to its long-term sustainability, it became the target of a hostile takeover. The first act of the person who took over the company was to clear-cut and liquidate the standing timber to pay off the junk bonds he had used to make the takeover possible, thus destroying the sustainability of the company and therefore its long-term monetary value, along with the job security of its employees.

If human society is to survive, however, the time has come to ask different questions: How much of any given resource is *necessary* to leave intact in Nature as a biological reinvestment in the health and continued productivity of the ecosystem for the benefit of both ourselves and the generations to come? How much of any given resource is *necessary* for us to use if we are to live in a reasonably comfortable lifestyle?

Necessity, in this sense, is a very different proposition from the collective societal "want, desire, need, demand" syndrome. If we are wise enough to curb our appetites and to embrace the concept of necessity instead of want, the Earth still has, we believe, enough resources to nurture us while we *re*learn how to nurture the Earth. Nurturing the Earth brings us to the concept of reversibility.

ARE THE CONSEQUENCES OF OUR DECISIONS REVERSIBLE, AND IF SO, TO WHAT EXTENT?

Sustainability potentially requires absolute reversibility in our decisions and subsequent actions if they prove ecologically and/or socially nonsustainable. Although the consequences of our decisions may be reversible in varying degrees in space and time, none are probably ever completely reversible. To understand this, let's consider a swamp.

If one wants to drain a swamp, one can dig a series of ditches which lead water away from the swamp and effectively lower the level of the water table. In so doing, one changes the habitat from a swamp to a meadow, a field, or something else, but only as long as the ditches remain functional.

If, however, one wants to reclaim the swamp, one must refill the ditches. While the water table will rise accordingly in the short term, it may take much longer for the life of the swamp to return. And because the swamp habitat has been set back to its beginning, it will never be as it would have been had the swamp not been drained. Nevertheless, a swamp has been regained, albeit a swamp somewhat different from the original one, and it can be protected over time should people choose to do so.

The notion that the consequences of our decisions are reversible only in degrees, but not in the absolute, raises such questions as: (1) How much reversibility is necessary to safeguard the sustainability of the future? (2) How much reversibility is acceptable to the community as a whole? (3) Who will make the decision and be accountable for the outcome? (4) What about the fate of things we introduce into our environment?

HOW WILL THE THINGS WE WANT TO INTRODUCE INTO OUR COMMUNITY'S ENVIRONMENT AFFECT ITS FUTURE?

We introduce thoughts, practices, substances, and technologies into the environment, and we usually think of those introductions in terms of development. Development in this sense is usually a commercial strategy to use or extract a given resource. Whatever we introduce into the environment, however, will consequently determine how the environment will respond to our presence and to our cultural necessities. It is therefore to the benefit of a community to pay close attention to what is introduced.

The introduction of a foreign substance, process, or technology has an impact on an ecosystem's ability to function. Consider the effects of some of the things we have introduced into the environment, because these things represent both our values and our behavior. Our initial introduction is our pattern of thought, which determines the way we perceive the Earth and the way we act toward it—either as something sacred to be nurtured or as only a commodity to be converted into money. Because our pattern of thought determines the value we place on various components of an ecosystem, it is our values that determine the way we treat those components and through them the ecosystem as a whole.

In our linear, product-oriented thinking, for example, an old-growth forest is an economic waste if its "conversion potential" is not realized; that is, the only value the old-growth trees have is their potential for being converted into money. Such notions stimulated Professor Garrett Hardin to observe that "Economics, the handmaiden of business, is daily concerned with 'discounting the future,' a mathematical operation that, under high rates of interest, has the effect of making the future beyond a very few years essentially disappear from rational calculation." Unfortunately, Hardin is correct. Conversion potential of resources counts so heavily because the economically effective horizon in most economic planning is only five years away. Thus, in our traditional linear economic thinking, any merchantable old tree that falls over and reinvests its biological capital into the soil is an "economic waste" because the potential of its wood fiber was not converted into money.

New equipment is therefore constantly being devised to make harvesting resources like trees ever more efficient. The chain saw, for example, greatly hastened the liquidation of old-growth forests worldwide. Possessed by this new tool, the timber industry and the forestry profession lost all sense of restraint and began cutting forests faster than they could grow. Further, no forested ecosystem has yet evolved to cope ecologically with the massive systematic and continuous clear-cutting made possible by the chain saw and the purely economic thinking behind it.

In our search for "national security" and cheap energy, we are introducing concentrated nuclear waste into many ecosystems, the impact of which is both global in scale and complex in the extreme. And there is no safe way to introduce the concentrations we are creating. The meltdown of the nuclear reactor at Chernobyl was not potentially as dangerous as the buried nuclear dump that blew up near Chelyabinsk, in the southern Ural Mountains, in late 1957 or early 1958. The land around Chelyabinsk was dead, and will be for perhaps centuries, over an area of roughly 1,000 square kilometers. All that was left standing after the explosion were chimneys of people's homes.

We have not the slightest idea how to deal safely with the concentrations of nuclear wastes we are introducing into the world. Consider the struggle in the United States to dispose of 52 tons of weapon's-grade plutonium ("the deadliest material known to man") left over from the Cold War.[65] After 24,000 years, only half of it will have decayed, turning into uranium 235, which is almost as deadly. "The point is," says Paul Levanthal, president of the Nuclear Control Institute, "[that] you can *never* get rid of it [emphasis added]."

The U.S. Department of Energy plans to turn two-thirds of the plutonium into mix-oxide, a fuel for nuclear reactors. Such a plan would create a security nightmare because this is the most vulnerable form of plutonium with respect to theft. In addition, it takes only 15 pounds of plutonium to make a nuclear bomb, and a skillful person could divert enough without sounding an alarm.[65]

Burying the remaining one-third of the tonnage of plutonium, which is what the Department of Energy wants to do, involves a process called vitrification. Vitrification entails mixing the plutonium with radioactive waste, fusing it with glass, and burying it far out of sight. Nevertheless, critics fear that the glass could break down in a thousand years or so and the plutonium could then accumulate, reach a critical mass, and set off a nuclear chain reaction.[65]

There are more immediate dangers, however, because the element is so dangerous that shavings of plutonium have been known to burst spontaneously into flames. In addition, the concrete bunkers built at the Rocky Flats Environmental Technology Site in Colorado have already developed cracks, and evidence of leaks has been measured in the adjacent land.[65]

Yet instead of committing our efforts to producing safe, clean solar and wind energy, we cling steadfastly to unsafe, dirty nuclear energy and create thousands of tons of nuclear waste annually through the military–industrial complex and peacetime technology. If we continue this course, the biosphere will eventually adapt to high, generalized concentrations of radioactivity, but most life as we know it will not be here to see that adaptation take place.

Our management of the resources, including local ones, maximizes the output of material products, putting the notion of "conversion potential" into operation. In so doing, we not only deplete the resource base but also produce unmanaged and unmanageable "by-products," often in the form of hazardous "wastes." In unforeseen ways, these "by-products" are altering the way our biosphere functions. In reality, there is no such thing as a "by-product;" there is only an unintended product, which more often than not is undesirable.

Because of unforeseen and usually undesirable effects from many of our introductions, we must shift our thinking from managing for particular short-term products to managing for a desired long-term condition on the landscape, an overall desired outcome of our decisions and actions.

As we come to recognize the potential for undesirable effects from the things we introduce into the environment, we must shift our thinking from being exclusively product oriented to becoming process oriented. We must become innovative and daring, and we must focus on controlling the type and amount of processes, substances, and technologies that we introduce into an ecosystem to effect a particular outcome. With prudence in our decisions about what to introduce into an ecosystem and how to do it, we can have an environment of desirable quality to support a chosen lifestyle, an environment that can still produce a good mix of products and amenities, but on an ecologically sustainable basis.

Looking ahead to all possible effects of what we introduce into our respective local environments is imperative. If we would ensure that any material introductions we make into the environment would be biodegradable as food for organisms like bacteria, fungi, and insects, then our "waste" would be their nutriment. In addition, if we use solar- and wind-based energy instead of fossil fuels, and if we recycle all nonrenewable resources in perpetuity, we will shift our pattern of thought from one that is ecologically exploitive to one that is ecologically friendly and sustainable.

Thus, in the name of sustainable community, the question becomes: How will the things we want to introduce into our community's environment affect our future and that of our children?

HOW MUCH WASTE CAN WE CONVERT INTO FOOD FOR MICROORGANISMS?

While we can manage what we introduce into the environment by the choices we make, once something is introduced, it is, in a vast majority of cases, out of our hands and our control. Consider, for example, the human-generated garbage from our throwaway society. How many garbage dumps, euphemistically called "sanitary landfills," are there now, and how many more will be needed for the next century?

Although much of what is discarded by our society may break down or be consumed by microorganisms over varying amounts of time, there is much that will undoubtedly be around for centuries. As the world's population continues to grow and towns and cities to swell with citizens and expand ever farther into the countryside, the human-generated garbage will increase proportionately while the available land area in which to hide it will continually shrink. This situation is untenable in the sense of social/environmental sustainability.

Paul Hawken puts it succinctly: "Industry has transformed civilization and created material wealth for many people.

But it only succeeds by generating massive amounts of waste. Our production systems function by taking resources, changing them to products and discarding the detritus back into the environment. This process is overwhelming the capacity of the environment to metabolize our waste, and, as a result, the health of our living systems is slowly grinding down. There are simply too many manufacturers making too much too fast....By the time you get to the consumer, 98 percent of the problem has already occurred."[66]

We must therefore specifically ask: How much human-generated waste can we convert into food for microorganisms so that it may continually cycle throughout the environment as a renewable source of energy? In addition, we must ask of each new item that technology proposes to produce—while it is still in the design stage: Will this item be biodegradable?

If the answer is yes, then we must ask: By what mechanism will it be biodegradable? To what degree will it be biodegradable? In what time frame? And under what conditions?

If the answer is no, then we must ask: Why is it not biodegradable? Can it be made to be biodegradable? If not, is there a biodegradable substitute that can serve the same purpose? If not, can one be made? If not—don't make it!

According to Paul Hawken, there is a growing movement (the Factor Ten "Club") in both Europe and the United States to radically reduce our requirements from resources. The Factor Ten "Club" consists of scientists, economists, and experts in public policy who are calling for a 90 percent reduction (factor ten) in the use of materials and energy over the coming 40 to 50 years.[66]

The goal of Factor Ten is to invent products, technologies, and systems that deliver the services people want without 90 percent of the fuel, metal, wood, packaging, and waste that is currently used. Keep in mind, says Hawken, that we want the service a product provides, not the thing itself. For example, we want clean clothes, not necessarily the detergent that cleans them.[66]

The upshot is that society can no longer afford to produce and use things that are wasteful and are not rapidly biodegradable. But unless these questions are specifically asked, new technology will most likely produce more and more things that are *non*biodegradable to increasing degrees. This does not mean that planned obsolescence as a corporate marketing strategy will disappear; it only means that whatever is discarded will take up more and more space in a nonusable form for ever-longer periods of time, like industrial chemicals, metals like aluminum, some plastics, and nuclear waste already do.

Nonbiodegradable materials, especially those that are toxic and capable of spreading, can make planning for social/environmental sustainability exceedingly difficult, even at the scale of a local community. The negative effects of such materials are often more poignant, however, at the bioregional scale, where they may affect an entire supply of water or air.

And speaking of nonbiodegradable materials, what about those of past decades that now lie buried in sanitary landfills? Is any of it reusable? Yes, according to the Berkshire County Council, in England, which is going to "mine" old landfills for reusable glass, metals, and plastics to become raw materials for the recycling business. This action (along with increasing present-day recycling and reduction of such things as flagrant overpackaging) will begin to clean up the environment, create jobs, and increase the life span of already established landfills.[67]

WILL PLANNING BENEFIT US AS A COMMUNITY?

Local people who empower themselves to work together in tapping the utmost powers of the mind, intuition, and experience in developing their own sustainable community will traditionally reap the following benefits:

1. A defined course of action (a vision), which helps ensure that the selected course has a good chance of success.
2. A process that serves as the foundation on which all community activities are based and, as such, must result in answers to what, where, when, why, and how actions are to be taken and who will conduct the actions for whom. If these questions are answered in a manner satisfactory to all, the chance of a destructive conflict is greatly reduced, but if perchance conflict arises, it can usually be resolved, often within the context of sustainable community development.
3. A well-conceived plan that allows those responsible to determine what they are responsible for and provides people with the opportunity to gain clear insight with respect to their specific tasks in relationship to the function of a community as a whole.
4. A process that helps a particular group of people communicate to others that the group is thoughtful in what it is doing and stands a good chance of accomplishing its stated purpose.
5. A process that will aid in monitoring and evaluating a community and its achievements.
6. A periodic evaluation of a group's progress toward meeting its vision, goals, and objectives identified in the plan, which is critical for evaluating whether it is providing the promised services to its customers and supporters. This step is essential for the health and growth of any community or organization within a community.
7. A vehicle through which the collective long-range (100+-year) vision of the people involved with a community can be realized. Planning, which is looking at options and solutions, helps people focus their energy on their vision, goals, and objectives and thereby helps a community achieve maximum utilization of its human talents and financial resources.

8. A process that helps people to influence what the present is and the future might be for the benefit of both today's citizens and tomorrow's generations.

In addition to all the goals, parameters, and legal requirements embedded in the planning process, it is fundamental that leaders endorse the concept of persons, which begins with recognition, understanding, and acceptance of people's diversity and intuition in their creative gifts, talents, and skills. These creative gifts, talents, and skills will also be needed to design lifestyles that are in sustainable harmony with the environment, which affects the sustainable harmony of the whole world.

Having said all of this, we are reminded of a thought expressed by a Chinese Mandarin to a Chinese Army major in the movie *Inn of the Sixth Happiness,* which is important to understand in the planning process, despite the fact that on the surface it seems contrary to planning: "A planned life, my friend, is a closed life. It can be tolerated perhaps, but it cannot be lived." To expand on this notion, we will borrow from an article by Kathy Gottberg, an ex-planner.[68]

As members of Western society, we are encouraged to plan as a response to the Newtonian model of a world characterized by materialism and reductionism. This mechanistic world view says in effect that if we can control all the pieces of our lives, we can plan for and thus control the outcome of our experiences, an idea that defines us as a cog within a huge machine.

As such, our highest goals are to plot our lives and circumstances to produce the least friction, to create a new and better operating system or a richer, more rewarding feedback loop. These ideas are dualistic, however, and serve to separate us from other people and from Nature.

This sense of separation creates an image of us as isolated individuals on our home planet, each trying to control events and circumstances in our moment-to-moment experiences, which produces an underlying feeling that life is a solitary battle we are losing. We seek security and our sense of identity in our individual accomplishments, rather than trusting an interconnected, interactive consciousness.

We therefore replace spontaneous creativity with a checklist of things to do, thus saying that we can't trust anyone else to look out for us, that we must look out for ourselves, which leaves us feeling alone and isolated in the universe, often with a palpable feeling of powerlessness. In addressing this lack of trust, Mahatma Gandhi once asked: "Does not the history of the world show that there would have been no romance in life if there had been no risks?" And yet, risk is what we are constantly trying to minimize or even eliminate.

Be that as it may, author Deepak Chopra says that every intention and desire has inherent within it the mechanics of its fulfillment. "Intention" in this sense is different than planning in that intention expresses the desire and then allows the spontaneous, intuitive outworking of the results. Planning, on the other hand, is the linear attachment to the detail and outcome, which often contradicts trust in the spontaneous and the intuitive, thus limiting possible outcomes.

If we don't trust our spontaneous creativity and our intuition, if we don't trust one another to act appropriately in an other-centered fashion, then we feel forced to "plan" an alternative, a contingency. But Nature offers another perspective.

In the Australian savannah are 20-foot-tall termite towers, which offer a magnificent demonstration of intuitive (instinctive in the case of termites) spontaneous creation and self-organization. These intricate engineering masterpieces, created from unplanned movement of many individual termites, are the tallest structures on earth relative to the size of their builders.

Although termites possess little individual physical capacity, they do have a strong sense of self and are constantly tuned into one another and their environment, which instinctively forms them into groups by attracting them to one another. They wander seemingly at will, bumping into one another and responding to the bump. When a critical mass of termites has gathered in a given location, their behavior shifts into an emerging action.

Their limited individual capabilities merge into a collective capacity and they begin building their towers. A group on one side begins building an arch. Another group notices this and begins constructing the other side of the arch in a spontaneous action that meets in the middle sans engineer, boss, or planner.

When first studying these termite towers, entomologists sought a planner, a leader. But after years of study, they concluded that the termite's towers are examples of "emergent properties," which means when a group is together, it is capable of behaviors that are simply unknowable when considering an isolated individual. Put differently, for all our study of the pieces of Nature, ourselves included, we will never see the potential for the collective possibility when considering the isolated individual.

There are many examples of emergent organizations, on both the micro and macro scale, and they have some things in common. First, there is a very strong sense of self and purpose in each individual participating in the collective. Individuals know who they are, collectively what they want to be or where they want to go (their vision), and they are tuned into themselves and to one another. Second, they recognize the value of quality relationship and the necessity of a free flow of information within the group or system. Third, each individual may wander at will, bump into others with attractive energy, and respond, which is contrary to most traditionally organized interactions because it cannot be planned.

The point of this discussion is to demonstrate that each of us is capable of infinitely more than the sum of our parts, far exceeding our expertise in planning. The quantum potential of our spontaneous creativity cannot be accessed by lists, calculations, or manipulations. It can only be accessed by a willingness to let something better than a rigid plan happen and participate intuitively in the happening. It is acting out of mature awareness, with full responsibility and participation in the dance of life.

To get the maximum benefit out of any plan, therefore, people must let creative spontaneity be part of the process. A plan must always be flexible and open to serendipitous opportunities, where the exact outcome may be uncertain, but if the direction intuitively feels right, then "trust in the result" must be the guiding principle.

"But how," you might ask, "do we know we are achieving what we desire?" That is the purpose of monitoring.

WHY MONITOR FOR SUSTAINABILITY?

Although the word "monitor" is variously construed, its meaning here is to scrutinize or check systematically with a view to collecting specific kinds of data that indicate whether one is moving in the direction one wishes. Monitor has the same origin as monition, which means a warning or caution, and is derived from the Latin *monitio,* a reminder.

With respect to social/environmental sustainability, monitoring means to keep watch over and warn in case of danger, such as straying from a desired course. Monitoring is to remind us of activities that we already know are too harsh and could offend the system; on the other hand, it is to help us conserve the options embodied within the system for ourselves and future generations, but this requires the ability to ask relevant questions because monitoring is dependent on questions.

Learning how to frame good and effective questions is paramount not only for crafting a collective vision for the future but also for the process of monitoring what is necessary to achieve the vision. A question is a powerful tool when used wisely, because questions open the door of possibility. For example, it was not possible to go to the moon until someone asked the question, "Is it possible to go to the moon?" At that moment, going to the moon became possible. To be effective, each question must: (1) have a specific purpose, (2) contain a single idea, (3) be clear in meaning, (4) stimulate thought, (5) require a definite answer to bring closure to the human relationship induced by the question, and (6) explicitly relate to previous information.

In a discussion about going to the moon, one might therefore ask, "Do you know what the moon is?" The specific purpose is to find out if one knows what the moon is. Knowledge of the moon is the single idea contained in the question. The meaning of the question is clear: Do you or do you not know what the moon is? The question stimulates thought about what the moon is and may spark an idea of how one relates to it; if not, that can be addressed in a second question. The question as asked requires a definite answer, and the question relates to previous information.

A question that focuses on "right" versus "wrong" is thus a hopeless exercise because it calls for human moral judgment, and that is not a valid question to ask of either an ecosystem or science. If, however, one were to ask if a proposed action was good or bad in terms of one's community's collective vision, that is a good question.

For example, a good short-term economic decision may simultaneously be a bad long-term ecological decision and thus a bad long-term economic decision. To find out, however, one must ask: Although this is a good short-term economic decision, is it also a good long-term ecological decision and hence a good long-term economic decision? An answer to anything is possible only when the question has been asked.

In essence, questions lead to the array of options from which one can choose. Conversely, without a question, one is blind to the options. Learning about the options is the purpose of monitoring. In turn, to know what to monitor and how to go about it, one must know what questions to ask because an answer is only meaningful if it is in response to the right question.

Good monitoring has five steps: (1) crafting a vision, goals, and objectives; (2) preliminary monitoring or inventory; (3) monitoring implementation; (4) monitoring effectiveness; and (5) monitoring to validate the outcome.

Step 1: Crafting a vision, goals, and objectives—Crafting a carefully worded vision and attendant goals and objectives that state clearly and concisely your desired future condition and how you propose to get there within some scale of time is the necessary first step in monitoring so that you know what you want, where you want to go, and what you think the journey will be like. The vision and its goals and their objectives form the context of the journey against which you measure (monitor) all decisions, actions, and consequences to see if in fact your journey is even possible as you imagined it and what the consequences of the journey might be.

Once you have completed your statements of vision, goals, and objectives, you not only will be able to but also must answer the following questions concisely: (1) What do I want? (2) Why do I want it? (3) Where do I want it? (4) When do I want it? (5) From whom do I want it? (6) How much (or how many) do I want? (7) For how long do I want it (or them)? If a component is missing, you may achieve your desire by default, but not by design.

Only when you can answer all of these questions concisely do you know where you want to go and the value of going there, and only then can you calculate the probability of arrival. Next you must determine the cost, make the commitment to bear it, and then commit yourself to keeping your commitment. Only then are you ready for the next steps in monitoring.

Step 2: Preliminary monitoring or inventory—Preliminary monitoring is to carefully observe and understand the circumstances with which one begins, which means taking "inventory" of what is available in the present. Taking inventory requires the following questions: What exists now, before anything is purposefully altered? What condition is it in, and what is its prognosis for the future? Even though preliminary monitoring may require multiple questions, the outcome is still a single realization.

If, for example, you go to your doctor for an annual checkup, the doctor not only would have to take a series of measurements, such as your blood pressure and blood tests, but also would have to know what a healthy person is (including, if possible, you as a healthy person) as a benchmark against which to assess your current condition. If you are indeed healthy, then all is well; if not, your doctor would presumably prescribe tests to pinpoint what is wrong and ultimately prescribe medicine to correct your ailment and make a prognosis for your future.

If you go to your doctor but only allow him or her to take your blood pressure without checking the level of your cholesterol, your doctor cannot deal with your health as a systemic whole, and thus loses the ability to see the various components as parts of an interactive, interdependent system. In this respect, your body is similar in principle and function to your family, the community in which your family resides, the landscape in which your community rests, and the landscape within the bioregion.

Step 3: Monitoring implementation—Monitoring the implementation of projects on the ground asks the following question: Did we do what we said we were going to do? Although this type of monitoring is really just documentation of what was done, it is critical documentation because without it, it may not be possible to figure out what went awry (if anything did), how or why it went awry, or how to remedy it. Thus, to continue the doctor analogy, it is important to document whether your doctor really did the test he or she deemed necessary because doubt as to the performance of one test can seriously obscure the results and failure to perform a test can alter completely (and perhaps disastrously) the doctor's interpretation of the outcome.

Step 4: Monitoring effectiveness—Monitoring to assess effectiveness means monitoring to assess the implementation of your objectives, not the goals or vision. Recall from our earlier discussion that a vision and its attendant goals describe the desired future condition for which you are aiming. They are qualitative and thus not designed to be quantified. An objective, on the other hand, is quantitative and so is specifically designed to be quantifiable.

Monitoring to assess effectiveness of an objective requires asking: Is the objective specific enough? Are the results clearly quantifiable and within specified scales of time? Monitoring the effectiveness of your project with the aid of indicators provides information (feedback) with which to assess whether you are in fact headed toward the attainment of your desired future condition (the condition of your collective vision), maintaining your current condition, or moving away from your desired future condition.

Monitoring for effectiveness means the systematic monitoring of indicators that are relevant to achieving your vision. A good indicator helps a community recognize potential problems and provides insight into possible solutions. What a community chooses to measure, how it chooses to measure it, and how it chooses to interpret the outcome will have a tremendous effect on the quality of life in the long term.

Indicators close the circle of action by both allowing and demanding that you come back to your beginning premise and ask (reflect on) whether, through your actions, you are better off now than when you started: if so, how; if not, why not; if not, can the situation be remedied; if so, how; if not, why not; and so on.

Here a caution is necessary. Traditional unidimensional indicators, which measure the health of one condition (say the economy), ignore the complex relationships among economy, community, environment, neighboring communities, and the bioregion. When each component is viewed as a separate issue and thus monitored in isolation, measurements tend to become skewed and lead to ineffective policies, which in turn can lead to a deteriorating quality of life. Indicators must therefore be multidimensional and must measure the quality of relationships among the components of the system being monitored if a community is to have any kind of accurate assessment of its sustainable well-being.

Only with relevant indicators and a systematic way of tracking them is it possible to make a prognosis for the future based on your vision, goals, and objectives, which state a desired future condition within some scale of time and a plan to achieve it. Only with relevant indicators and a systematic way of tracking them is it possible to make the necessary target corrections to achieve your vision because only then can you know which corrections to make.

Returning to the doctor analogy, assessing effectiveness asks: Was the right test used (was it relevant to your condition)? Was the test effective (if it was the right test, did it perform as it was supposed to)? These are important questions, because if the wrong test was used or the test was ineffective, the results can be very different from those you were led to expect and the outcome could be unexpectedly life-threatening.

Step 5: Monitoring to validate the outcome—Monitoring for validation of the outcome is considered by many to be research. This type of monitoring involves testing the assumptions that went into the development of your objectives and the models on which they are based.

Monitoring for validation may require asking such questions as: Why didn't the results come out as expected? What does this mean with respect to our conceptual model of how we think the system works versus how the system actually works? Will altering our approach make any difference in the outcome? If not, why not? If so, how and why? What target corrections do we need to make to bring our model in line with how the system really works?

Validation is a necessary component of any monitoring plan because this is where you learn about the array of possible target corrections. In addition, monitoring for validation may have wide application for other projects.

Visiting once again the doctor analogy, suppose your doctor says, "The results of your tests show that you may have an unusual form of 'Highfalutin Disease.' Let me check with my colleagues and the literature to see what is known about it."

Three weeks later, your doctor calls you into the office and says, "Your particular form of 'Highfalutin Disease' is indeed unusual, and I can find no common cure. But there is a drug called 'Dumpin Highfalutin,' which shows promise in laboratory tests. It is now ready for human trials to see how well they corroborate laboratory results. If you are willing to become part of a controlled medical experiment, I think you may qualify for the drug. It's your best chance at the moment to regain your health."

In the end, it is through the questions we ask that we derive our vision, goals, and objectives. It is the questions we ask that frame our perceptions, that direct our actions, which require monitoring. And it is the questions we ask that determine what and how we monitor. Therefore, social/environmental sustainability depends first and foremost on the relevancy of the questions we ask, for they become our compass and map into the future.

SUMMARY AND CONCLUSION

What have we learned in our journey through this book? We learned something about community, conflict, communication, economics, relationship, governance, shared vision, and the questions we need to ask. Let's recap briefly.

Community—A community grows out of people coming together in a particular place to which they learn to feel a sense of fidelity. Over time, a community's fidelity to its place in the landscape evolves into a historical relationship of reciprocity in which the people change the landscape by living in it and interacting with it even as the landscape alters the people's image of themselves and their community.

Further, the existence of a community is based on the mutual trust of its inhabitants for one another, even to the point of accepting a scoundrel who can be reliably counted on to be a scoundrel. Finally, the notion of community is confined, except metaphorically, to the local residents, which means that local communities are the basic building blocks of social/environmental sustainability, albeit conflict arises from time to time.

Conflict—Destructive conflicts are best resolved through the transformation of individual consciousness, which increases people's compassion for one another and deepens one's understanding of the personal fears and struggles that we each daily face. Resolving conflicts through the transformative approach is critical because it emphasizes the capacity for personal growth embodied in the ability to accept risk. It also helps parties empower themselves to define the issues and decide the settlement in their own terms and in their own time through a better understanding of one another's perspectives. It therefore transforms society for the better by bringing out the intrinsic good in people, often through their willingness to compromise, which is based on good communication.

Communication—Although most communication is conveyed in tone of voice, body language, attitudes, vibrations, and other energies, we are concerned in this book primarily with the spoken word. Words are symbols for the things we experience; therefore, the more accurately a chosen word builds a bridge to our common ground, the easier it is to get in touch with one another, stay in touch, build trust, and ask for and receive help.

Emotions and knowledge (a reflection of social experience) are shared through communication, which is perhaps one of the most difficult things we do as human beings, and yet it is simultaneously one of the most important things we do. We are creatures who must share feelings, senses, abstractions, and concrete experiences in order to know and value our existence in relation with one another. Communication is the way we share the essence of our relationships. Our very existence revolves around it, and without it, we have nothing of value.

Language as a way of sharing and coordinating our human values is becoming increasingly important and complex as persons of different languages and ethnic backgrounds move into communities or visit them seasonally to share the harvest of renewable natural resources from their immediate landscapes. It is therefore necessary in considering sustainable community development to address language as a means of communication, including cross-cultural communication, which in turn is necessary to understand how different people view economics.

Economics—For our discussion of economics to end up with holistic and connected conclusions, we had to start with some linear and hierarchical definitions. A true revolutionary would have noted the irony of using tools from an old order to establish a new one, such as a careful assessment of the phenomenon of scarcity, which must become the centerpiece of economic thinking and methodology if we are to harness economic growth in the interest of achieving social/environmental sustainability.

Clarity in understanding the language of economics, such as the meaning of "scarcity," is all-important, as we have learned, if productive and concrete steps are to be taken, because economics and sustainability do not appear to be comfortable bedfellows. After all, the historical promotion of growth has occurred not only because economists have applied their formal disciplinary tools but also because the many and diverse people dealing with business in both the public and private sectors have been equally ardent in championing unbridled economic growth. The attempts to harness economic jargon in the service of a particular vested interest or point of view have been nothing short of monumental.

It is therefore important to realize that both ecology and economy have the same Greek root, *oikos,* a house. Ecology is the knowledge or understanding of the house and economy is the management of the house, and *it is the same house.* And a house divided against itself cannot long stand. Yet, it has been the assumption of our society that if we manage the parts correctly, the whole will come right. Although evidence to the contrary now comes from all directions, our systems of knowledge and management are still structured around this assumption.

"Call a thing immoral or ugly, soul-destroying or a degradation of man, a peril to the peace of the world or to the well-being of future generations," wrote E.F. Schumacher, "as long as you have not shown it to be 'uneconomic' you have not really questioned its right to exist, grow, and prosper." This quote points out the untenable divisiveness that weakens the house inhabited by ecology and economy. The question, therefore, is how to heal the current division in the house.

A major problem with addressing (and, if necessary, suggesting adjustment to) the role of economic philosophy and theory is the connection between environment and economics. Be assured that the loss of biodiversity, pollution of the Pacific Ocean, and global warming are all directly caused by human economic activity—activity that, we will be reminded in no uncertain terms, is designed to feed and house the peoples of the world and in a number of ways to make life better for us all.

Here we confront a major irony in the consideration of our growth-oriented economy and public policies: The activities promoted as improving the well-being of people end up being directly and indirectly responsible for the degradation of natural and human-made systems. These activities thus cause considerable loss of human satisfaction, both current and future. So we end up degrading human well-being and maybe even seriously threatening the global carrying capacity of the human population as we energetically and unconsciously promote human well-being in economic isolation of the ecological systems on which that well-being depends.

The really important questions about the theory and practice of economics are those concerning human values and distribution of goods and services. "Who gets what?" is the only important question, since "How much can we produce?" (a question essentially for engineering and technology) will inevitably lead to an answer of "not enough." That answer assumes, of course, that the realities of finite resources are not reconciled with the impossibility of the "more-is-better" mindset about the distribution of goods and services encouraged by the Western industrial ethic of continual economic growth.

"Who gets what?" is important not only in terms of economic theory and practice but also in terms of human relationships. What economic theory and practice ignore is the *quality* of interpersonal human relationships. This is a tragic oversight because it is the quality of human relationships in all their myriad forms that not only gives life its value but also makes sustainable community development a real possibility.

Relationships—Relationships are the strands in the web of life, and there is no escaping the web. Relationship is all there is. In fact, life is relationship, and relationship is life. One cannot exist without the other because Nature's design is a continual flow of cause-and-effect relationships that precisely fit into one another at differing scales of space and time and are constantly changing within and among those scales. Every day of our lives is therefore about learning how to relate to ourselves, one another (both present and future), and our environment. In short, our entire lives are spent learning how to be successful in an ever-expanding array of relationships, including social governance itself.

Governance—Democracy is the backbone of local community development and ultimately community sustainability, and like sustainable community development, governance is an evolutionary process. And because every phase of organizational growth and development within a community is determined by its membership, so too must the appropriate forms of democratic governance (structure/process) be determined by the membership.

Democracy is a system of shared power with checks and balances, a system in which individuals can affect the outcome of political decisions. Within this system, we can make real and profound changes. All that is required is education, a purposeful vision, and political will, which means having the courage to come together and to choose to act, the knowledge to know how to act, and the vision to know where to go.

Democracy is designed to protect individual freedoms within socially acceptable relationships with other people individually and collectively. People practice democracy by managing social processes themselves. Democracy is another word for the responsibility of self-directed social evolution.

Democracy in the United States is built on the concept of inner truth, which in practice is a tenuous balance between spirituality and materialism, intuition and knowledge. One such truth is the notion of human equality, in which all people are pledged to defend the rights of each person, and each person is pledged to defend the rights of all people. In practice, however, the whole endeavors to protect the rights of the individual, while the individual is pledged to obey the *will* of the majority, which may or may not be just to each person because the majority may not be just. And when the majority is unjust, it should not win. In the end, the best chance for justice lies in a shared vision of the future toward which to build.

Shared vision—A vision consists of what people ideally want, not of what they fear. A shared vision of a sustainable future toward which a community can build creates confidence, consensus, and energy in equal parts. At a deeper level, it engages our imagination and helps to ferret out which questions need to be asked, how to word them, and when to ask them.

By engaging our imagination and our sense of possibility of the ideal through countless small-scale initiatives, such as a shared community vision, we the people can create an opportunity to confirm a more positive sustainable future with a healthy environment and a just society. To do so, however, we must change our values and habits if we are to be ecologically sustainable and socially inclusive in our ways of living. A shared vision of the future starts the process of change by enabling people to think differently about their lives and in so doing commences to change them.

A resident community, through its vision, accepts responsibility for its own survival, and outsiders must fit themselves into that vision if the community is to be sustainable. A vision can thus be likened to a human body in that the strong organizing context of the body's division of labor keeps the various cells functioning within acceptable bounds. Hence a community's need for the strong organizing context of a vision.

The questions we need to ask—Because the old self-centered questions and old self-centered answers have led us into today's disintegration of social/environmental integrity and are leading us toward an even greater loss of integrity tomorrow, it is important to understand that the answer to a problem is only as good as the question and the means used to derive the answer. Before we can arrive at fundamentally new answers, we must be willing to risk asking fundamentally new questions. We must therefore pay particular attention to the questions we ask because the answers we accept will become the consequences of the future. This means that we must look long and hard at where we are headed with respect to the quality of our community, our environment, and the legacy we are leaving the children.

Heretofore our society has been more concerned with *getting* politically correct answers to its self-centered questions than it has been with *asking* other-centered (morally right) questions. Politically correct answers validate preconceived economic/political desires. Other-centered questions lead us toward a future in which social/environmental options are left open so that generations to come may define their own ideas of a "quality community" and a "quality environment" from an array of possibilities.

To this end, we must pay vastly more attention to the questions we ask. It is, after all, the questions we ask that guide the sustainability of our community/environmental development, and it is the questions we ask that determine the options we bequeath to the future. Asking the right questions, therefore, can create a holistic web of multilevel thinking, which can act as a catalyst for conscious responses to the complexity of human interactions within the organic system of nature, including the community itself.

Although it was not within the scope of this chapter to offer an exhaustive list of questions that need to be asked, we discussed ten questions the answers to which we feel will determine whether a vision for the future is sustainable: (1) What sources of energy are available to our community? (2) What social capital is available within our community? (3) How can specialities be sustainably fit into a general community? (4) What is necessary to build a community in an intelligent, moral way? (5) When is enough, enough? (6) Are the consequences of our decisions reversible, and if so, to what extent? (7) How will the things we want to introduce into our community's environment affect its future? (8) How much waste can we convert into food for microorganisms? (9) How will planning benefit us as a community? (10) Why monitor for sustainability?

For our respective hometowns—and yours—to enjoy social/environmental sustainability, their citizens must find out what worked in the past and begin recreating it in the present, and where problems arise (as they will) work together to resolve them, which means accepting the risk of trying new things for a future that is fast approaching. The only way to create, maintain, and pass forward a sense of community is by working together, because the trust and friendliness of a community are founded on the quality of its interpersonal relationships, of which we are all an integral part. Ultimately, it is the quality of the human relationships within our respective communities—present and future—to which we must all be committed if social/environmental sustainability is to become a reality.

ENDNOTES

1. Chris Maser. 1996. *Resolving Environmental Conflict: Towards Sustainable Community Development*. St. Lucie Press, Boca Raton, FL.
2. Chris Maser. 1994. *Sustainable Forestry: Philosophy, Science, and Economics*. St. Lucie Press, Boca Raton, FL.
3. The Associated Press. 1987. Whistle blower's actions branded disloyal by Army. *The Oregonian* (Portland, OR) November 10.
4. Georg Feuerstein, Subhash Kak, and David Frawley. 1995. The Vedas and perennial wisdom. *The Quest* 8(4):32–39, 80–81.
5. Wendell Berry. 1993. *Sex, Economy, Freedom, and Community*. Pantheon Books, New York.
6. Calvin S. Hall, Gardner Lindzey, John C. Loehlin, and Martin Manosevitz. 1985. *Introduction to Theories of Personality*. John Wiley & Sons, New York.
7. Ronald L. Warren. 1972. *The Community in America* (2nd ed.). Rand McNally College Publishing, Chicago.
8. Joanne Jacobs. 1996. Americans' civic involvement still strong, but changing. *Corvallis Gazette-Times* (Corvallis, OR) July 25.
9. Abraham Maslow. 1985. *The Farther Reaches of Human Nature*. Penguin Books, New York.
10. M. Boyd Wilcox. 1996. Residents should have right to cap population growth. *Corvallis Gazette-Times* (Corvallis, OR) September 10.
11. For an in-depth discussion of transformative facilitation, see Chris Maser. 1996. *Resolving Environmental Conflict: Towards Sustainable Community Development*. St. Lucie Press, Boca Raton, FL; Robert A. Bush and Joseph P. Folger. 1994. *The Promise of Mediation*. Jossey-Bass, San Francisco.
12. The discussion of communication is based in large measure on the Aviation Instructor's Handbook. 1977. Federal Aviation Administration, U.S. Department of Transportation, U.S. Government Printing Office, Washington, D.C.
13. James Allen. 1981. *As a Man Thinketh*. Grosset & Dunlap, New York.
14. Don Costar. 1995. No need to change grazing practices. Letters to the editor. *Reno-Gazette Journal* (Reno, NV) January 13.

15. The discussion of words is based on Communication and Leadership Program. 1995. Toastmasters International, Mission Viejo, CA, as well as the experience of the senior author.
16. The discussion of foreign views of Americans is based in part on James G. Patterson. 1996. Communicating and negotiating internationally. *The Toastmaster* 62(9):9–10 and Desmond Morris. 1996. Louder than words. *The Toastmaster* 62(9):11–12, as well as the experience of the senior author.
17. Daniel J. Boorstin. 1983. *The Discoverers: A History of Man's Search to Know His World and Himself.* Vintage Books, New York.
18. The discussion of body language follows Desmond Morris. 1996. Louder than words. *The Toastmaster* 62:11–12, as well as the experience of the senior author.
19. The discussion in this section is based on Victor M. Parachin. 1996. 10 myths about people with disabilities. *The Toastmaster* 62(11):24–26 and Dot Nary. 1996. Keeping the "dis" out of disable. *The Toastmaster* 62(11):27–28.
20. Chris Maser. 1997. *Sustainable Community Development: Principles and Concepts.* St. Lucie Press, Boca Raton, FL.
21. Donald Ludwig, Ray Hilborn, and Carl Walters. 1993. Uncertainty, resource exploitation, and conservation: lesson from history. *Science* 260:17, 36; Chris Maser. 1992. *Global Imperative: Harmonizing Culture and Nature.* Stillpoint Publishing, Walpole, NH.
22. Donella H. Meadows, Dennis L. Meadows, Jørgen Randers, and William W. Behrens. 1972. *Limits to Growth.* Universe Books, New York.
23. Carl G. Jung. 1958. *The Undiscovered Self.* A Mentor Book, New York.
24. Per Bak and Kan Chen. 1991. Self-organizing criticality. *Scientific American* pp. 46–53.
25. William K. Stevens. 1990. New eye on nature: the real constant is eternal turmoil, *The New York Times* July 31.
26. John J. Magnuson. 1990. Long-term ecological research and the invisible present. *BioScience* 40:495–501.
27. Fritjof Capra. 1975. *The Tao of Physics.* Shambhala, Berkeley, CA.
28. Discussion of Silver City's water catchment is from J.T. Columbus. 1980. Watershed abuse—the effect on a town. *Rangelands* 2:148–150.
29. Wally W. Covington and M.M. Moore. 1991. Changes in Forest Conditions and Multiresource Yields from Ponderosa Pine Forests Since European Settlement. Unpublished report submitted to J. Keane, Water Resources Operations, Salt River Project, Phoenix, AZ, 50 pp.
30. Peter Berg and Raymond Dasmann. 1978. Reinhabiting California. pp. 217–220. *In:* Peter Berg and Raymond Dasmann (Eds.). *Reinhabiting a Separate Country: A Bioregional Anthology of Northern California.* Planet Drum, San Francisco.

31. Neil Jumonville. 1995. Most Americans now shun founder's radical doctrine. *Corvallis Gazette-Times* (Corvallis, OR) July 4.
32. Anna F. Lemkow. 1994. Our common journey toward freedom. *The Quest* 7(1):55–63.
33. Kevin Phillips. 1994. *Arrogant Capital: Washington, Wall Street, and the Frustration of American Politics.* Little, Brown, New York.
34. The quote by Alexander Tayler was given to Chris by a friend who cannot remember where he got it, and we have found nothing on Alexander Tayler in the literature. It is included because it rings true, truer than the vast majority of Americans would probably care to admit.
35. Ed Klophenstein. 1996. New land-use law may limit public comment. *Corvallis Gazette-Times* (Corvallis, OR) March 26.
36. Charles Darwin. 1872. *The Expression of Emotions in Man and Animals.* Julian Friedmann Publishers, London.
37. Aldo Leopold. 1949. *A Sand County Almanac and Sketches Here and There.* Oxford University Press, New York.
38. Elisabeth Kübler-Ross. 1969. *On Death and Dying.* Macmillan, New York.
39. Michael Soulé. 1990. The onslaught of alien species and other challenges in the coming decades. *Conservation Biology* 4:233–239.
40. Phyllis Windle. 1992. The ecology of grief. *BioScience* 42:363–366.
41. P.E. Irion. 1990. Funeral. pp. 450–453. *In:* R.J. Hunter, H.N. Maloney, L.O. Mills, and J. Patton (Eds.). *Dictionary of Pastoral Care and Counseling.* Abingdon Press, Nashville, TN; P.E. Irion. 1990. Mourning customs and rituals. pp. 450–453. *In:* R.J. Hunter, H.N. Maloney, L.O. Mills, and J. Patton (Eds.). *Dictionary of Pastoral Care and Counseling.* Abingdon Press, Nashville, TN.
42. Colin Murray Parkes. 1974. *Bereavement.* International Universities Press, New York.
43. Ed Mayo and Edward Hill. 1996. Shared vision. *Resurgence* 177:12–14.
44. D.T. Suzuki. 1959. *Zen and Japanese Culture.* Princeton University Press, Princeton, NJ.
45. Lin Yutang. 1938. *Wisdom of Confucius.* Random House, New York.
46. The discussion of values and aspects of vision follows Laurence G. Boldt. 1993. *Zen and the Art of Making a Living.* Penguin/Arkana, New York.
47. Kevin Preister. 1996. Community assessment. *Community Ecology, A Newsletter of the Rogue Institute for Ecology and Economy* 2(1):1, 5.
48. Lewis Carroll. 1933. *Alice's Adventures in Wonderland.* Doubleday, Doran, & Co., New York.
49. *The World Book Encyclopedia.* 1985. World Book, Inc., Chicago.
50. Coldstream, Scotland. 1996. Stone of Scone returned. *Corvallis Gazette-Times* (Corvallis, OR) November 16.
51. David Brisco. 1997. Report: world still in bad shape. *Corvallis Gazette-Times* (Corvallis, OR) January 12.

52. Carol Zimmerman. 1997. Dear Mr. President. *Parade Magazine* January 12:10–11.
53. John H. Baldwin. 1984. *Environmental Planning and Management.* Westview Press, Boulder, CO, 280 pp.; Dr. Kenneth S. Krane, Chair, Department of Physics, Oregon State University, Corvallis (personal communication).
54. The discussion of talents follows Laurence G. Boldt. 1993. *Zen and the Art of Making a Living.* Penguin/Arkana, New York.
55. John Addington Symonds. no date. *The Life of Michelangelo.* Carlton House, New York.
56. The discussion in this section is based on Wade Chabassol. 1996. Inspiring officer performance. *The Toastmaster* 62(8):13–14.
57. Manfred Stanley. 1983. The mystery of the commons: on the indispensability of civil rhetoric. *Social Research* 50:851–883.
58. Mary Douglas. 1986. *How Institutions Think.* Syracuse University Press, Syracuse, NY.
59. Vinoba Bhave. 1994. Moved by love. *Resurgence* 165:26–27.
60. The discussion of study circles is based on Cecile Andrews. 1992. Study circles: schools for life. *In Context* 33:22–25.
61. Gifford Pinchot and Elizabeth Pinchot. 1994. Beyond bureaucracy. *Business Ethics* 8(2).
62. Elizabeth Pinchot and Gifford Pinchot. 1994. *The End of Bureaucracy and the Rise of the Intelligent Organization.* Berrett-Koehler, San Francisco; Gifford Pinchot and Elizabeth Pinchot. 1993. Unleashing intelligence. *Executive Excellence* September:7–8; Elizabeth Pinchot. 1992. Can we afford ethics? *Executive Excellence* March:1–2; Elizabeth S. Pinchot. 1992. Balance the power. *Executive Excellence* September:3–4; Gifford Pinchot. 1992. Rewarding with status. *Executive Excellence* August:3–5.
63. Austine Roberts. 1977. *Centered on Christ: An Introduction to Monastic Profession.* St. Bede's Publications, Still Rive, MA, 169 pp.
64. Joseph Campbell. 1968. *The Hero with a Thousand Faces.* Bollingen Series, Princeton University Press, Princeton, NJ, 416 pp.
65. Knight-Ridder Tribune News Service. 1996. U.S. struggles to get rid of plutonium. *Corvallis Gazette-Times* (Corvallis, OR) December 22.
66. Paul Hawken. 1996. Undoing the damage. *Vegetarian Times* September:73–79.
67. Lorna Howarth. 1997. Wealth in waste. *Resurgence* 180:23.
68. Kathy Gottberg. 1996. Confessions of an ex-planner. *The Quest* 9(4):12–14.

APPENDIX: SUSTAINABLE COMMUNITY RESOURCES*

The following list of resources is provided to support your continued investigation into and learning about efforts to create sustainable communities. This list is by no means exhaustive, but it will provide access to a network of organizations and people working to sustain communities within sustainable environments.

Adirondack Council
P.O. Box D-2, Elizabethtown, NY 12940
(518) 873-2240 adkcouncil@aol.com
Working to protect the natural environment and sustain human communities.

Aspen Institute—Rural Economic Policy Program
1333 New Hampshire Avenue NW, Suite 1070, Washington, D.C. 20036
(202) 736-5800 hn0435@handset.org
Fosters collaborative approaches to learning and economic development.

Black Range RC&D, Inc.
2610 North Silver, Silver City, NM 88061
(505) 388-9566
Assists local people in developing long-range programs for resource conservation and development.

Center for Rural Affairs
P.O. Box 406, Walthill, NE 68067
(402) 846-5428
Working to build sustainable rural communities through social and economic justice with environmental stewardship.

* The list of resources is provided for your information; it is not meant as an endorsement for any organization.

Chehalis Basin Fisheries Task Force
2109 Sumner, Suite 202, Aberdeen, WA 98520
(360) 533-1766
Diverse interests seeking to enhance fisheries resources through community outreach and education.

Coalition for a Livable Washington
2111 East Union, Seattle, WA 98122
(206) 324-3628 instwafut@aol.com
Works to link environmental and labor groups by promoting sustainable alternatives.

Colonial Craft
2772 Fairview Avenue North, St. Paul, MN 55113
(612) 631-3110
Manufacturer of wood products derived exclusively from certified sustainably grown wood.

Columbia-Pacific RC&D
303 South "I" Street, Aberdeen, WA 98520
(360) 533-4648
Educational resources for community sustainability.

Ecoforestry Institute
785 Barton Road, Glendale, OR 97442
(541) 832-2785
Dedicated to teaching and certifying holistic, ecologically sound forestry practices that protect and restore the sustainability of forests while harvesting forest products.

ECONorthwest
99 W. 10th Avenue, Suite 400, Eugene, OR 97401
info@eugene.econw.com
Policy analysis of allocation and use of natural resources, including jobs, land-use patterns, incomes, and industrial structure.

EcoTimber International
1020 Heinz Avenue, Berkeley, CA 94710
(510) 549-3000 ecotimber@igc.apc.org
Founded on the premise that the world's forests are most likely to be used wisely if they are managed sustainably and that the best way to promote sustainable forestry is through the marketplace.

Ecotrust
1200 NW Front Avenue, Suite 470, Portland, OR 97209
(503) 227-6225 info@ecotrust.org
Promotes conservation-based development through helping local communities

develop the capacity to meet human needs and maintain ecological integrity. Helps communities develop financial capital for sustainability.

Environmental Defense Fund
257 Park Avenue South, New York, NY 10010
(212) 505-2100 members@edf.org
Environmental advocacy linking science, economics, and law.

Flathead Economic Policy Center
15 Depot Park, Kalispel, MT 59901
(406) 756-8584
Addresses needs for proactive, community-based solutions to social, ecological, and economic concerns by promoting community trust and collaborative processes.

Forest Care, Inc.
437 Walnut Street, Statesville, NC 28677
(704) 873-5344
Private, for-profit consulting firm specializing in rural and urban sustainable forest management.

Forest Trust
P.O. Box 519, Santa Fe, NM 87504
(505) 983-8992
Assists rural communities of northern New Mexico with the creation of sustainable economic opportunities based on forestry and forest products.

Hoopa Tribal Forestry
P.O. Box 368, Hoopa, CA 95546
(916) 625-4284
Experience in practical ecosystem management in which retention of critical wildlife habitat is protected.

Institute for Sustainable Forestry
P.O. Box 1580, Redway, CA 95560
(707) 923-4719
Sustainable forestry wood certification.

Montana Women's Economic Development Group
127 N. Higgins, Missoula, MT 59802
(406) 543-3550
Working to sustain communities and create jobs that provide adequate and equitable standards of living for women and low-income people.

National Wildlife Federation
Northeast Natural Resources Center, 58 State Street, Montpelier, VT 05602
(802) 229-0650
Educates and inspires individuals and organizations to conserve natural resources for a peaceful, equitable, and sustainable future.

Natural Resources Defense Council
40 W. 20th Street, New York, NY 10011
(212) 727-2700 nrdcnyurban@igc.apc.org
Environmental advocacy.

Nez Perce Tribal Forestry Program
P.O. Box 365, Lapwai, ID 83540
(208) 843-7328
Applying a policy of multiple use and sustainable yield to manage tribal forestry resources.

Northwest Center for Progressive Research
P.O. Box 10272, Olympia, WA 98502
(360) 352-9833 nwcpr@halcyon.com
Supports progressive ideals for public policy that reflect accountability, responsible citizenship, and commitment to the common good.

Northwest Environmental Watch
1402 Third Avenue, Suite 1127, Seattle, WA 98101
(206) 447-1880 nwwatch@igc.apc.org
Monitors the environmental conditions of the Pacific Northwest through research on sustainable development and provides tools for reconciling people and place, economy and ecology.

Northwest Policy Center
327 Parrington Hall, Box 35360, University of Washington, Seattle, WA 98195
(206) 543-7900 npcbox@u.washington.edu
Source of information for rural development strategies, techniques to assist small businesses, and work force training. Develops policy research and alternatives.

Oregon Economic Development Department
775 Summer Street NE, Salem, OR 97310
(503) 986-0123
State agency supporting local community development.

Pacific Forest Trust
P.O. Box 858, Boonville, CA 95415
(707) 895-2090
Dedicated to the restoration, enhancement, and conservation of privately owned forests through practical, on-the-ground application of stewardship forestry.

Pacific Rivers Council
P.O. Box 10798, Eugene, OR 97440
(541) 345-0119 pacificriver@igs.apc.org
River conservation through development of scientific tools, public policies, and strategies for community development.

Pinchot Institute for Conservation
1616 "P" Street NW, Washington, D.C. 20036
(202) 797-6582 102635.2117@compuserve.com
Forestry and research on community forestry.

Plumas Corporation
P.O. Box 3880, Quincy, CA 95971
(916) 283-3739
Concentrating on local economic diversification through restoration of degraded riparian habitat and upland forest ecosystems.

Quinault Tribal Offices
P.O. Box 198, Taholah, WA 98586
(360) 276-8211
Working on development issues surrounding tribal resources, such as fisheries, forests, and water.

Rocky Mountain Institute
1739 Snowmass Creek Road, Snowmass, CO 81654
(970) 927-3851
A foundation dedicated to research and education that foster sustainable and efficient use of resources by uniting environmental and corporate groups.

The Rogue Institute for Ecology and Economy
762 A Street, Ashland, OR 97520
(541) 482-6031
Conducts research and helps communities organize themselves around sustainable natural resources. Forest Products Certification Program. Member of Collaborative Working Circle working with community-based forestry organizations to exchange experiences, facilitate peer training, and share ideas.

Rural Development Initiatives, Inc.
P.O. Box 265, Lowell, OR 97452
(541) 937-8344
Development of strategic planning and leadership for rural communities.

Shasta-Tehama Bioregional Council
P.O. Box 492036, Redding, CA 96049
Coalition working for bioregional sustainability.

Shorebank Corporation
7054 S. Jeffrey, Chicago, IL 60649
(312) 288-1000
Bank of assisting for-profit and nonprofit strategies for community development.

ShoreTrust Trading Group
Port of Ilwaco, P.O. Box 826, Ilwaco, WA 98624
A joint venture between Ecotrust and Shorebank that uses conservation-based

principles of development to guide its lending practices to assist local business with such things as loans and new marketing strategies.

Sonoran Institute
7290 E. Broadway, Suite M, Tucson, AZ 85710
(520) 290-0828 soninst@azstarnet.com
Promotes community-based strategies that protect the ecological integrity of the land while meeting the economic needs of landowners and communities.

Sustainable America
350 Fifth Avenue, Room 3112, New York, NY 10118
(212) 239-4221 sustamer@igc.apc.org
National coalition working to develop the capacity of member groups to engage in strategies and advancing public policies for sustainable economic development.

Sustainable Northwest
1200 Northwest Front Avenue, Suite 280, Portland, OR 97209
(503) 221-6911 SustNW@teleport.com
Working to strengthen the capacity of local communities to promote environmentally sound economic development in the Pacific Northwest.

Sustainable Seattle
c/o Metrocenter YMCA, 909 Fourth Avenue, Seattle, WA 98104
sustsea@halcyon.com
A volunteer network seeking to sustain the long-term cultural, economic, and environmental and social health and vitality of the Pacific Northwest.

The Nature Conservancy
7 E. Market Street, Suite 210, Leesburg, VA 22075
(703) 779-1728 bweeds@cced.org
Mission is to conserve plants, animals, and natural communities by protecting lands and water while assisting in the viability of local economics.

Watershed Research and Training Center
P.O. Box 356, Hayfork, CA 96041
(916) 628-4206
Promotes sustainable local economies within a sustainable forest ecosystem through research, education, and economic development.

Western Network
616 Don Gaspar, Sante Fe, NM 87501
(505) 982-9805
Assists communities in resolving disputes.

Willapa Alliance
P.O. Box 278, South Bend, WA 98578
(360) 875-5195 willapanet@igc.apc.org

Supports programs of education, research, resource management, and economic development to maintain the health of natural and human communities.

World Wildlife Fund
1250 24th Street, NW, Washington, D.C. 20037
(202) 293-4800
Works to protect endangered wildlife and habitats.

Yellow Wood Associates, Inc.
95 South Main Street, St. Albans, VT 05478
(802) 524-6141 hn4402@handsnet.org
Consulting firm that provides services in rural community and economic development, including training programs in community-based measurement and progress toward community goals.